The Connected Mathematics Project Staff

Project Directors

James T. Fey
University of Maryland

William M. Fitzgerald
Michigan State University

Susan N. Friel
University of North Carolina at Chapel Hill

Glenda Lappan
Michigan State University

Elizabeth Difanis Phillips
Michigan State University

Project Manager

Kathy Burgis
Michigan State University

Technical Coordinator

Judith Martus Miller
Michigan State University

Collaborating Teachers/Writers

Mary K. Bouck
Portland, Michigan

Jacqueline Stewart
Okemos, Michigan

Curriculum Development Consultants

David Ben-Chaim
Weizmann Institute

Alex Friedlander
Weizmann Institute

Eleanor Geiger
University of Maryland

Jane Mitchell
University of North Carolina at Chapel Hill

Anthony D. Rickard
Alma College

Evaluation Team

Diane V. Lambdin
Indiana University

Sandra K. Wilcox
Michigan State University

Judith S. Zawojewski
National-Louis University

Graduate Assistants

Scott J. Baldridge
Michigan State University

Angie S. Eshelman
Michigan State University

M. Faaiz Gierdien
Michigan State University

Jane M. Keiser
Indiana University

Angela S. Krebs
Michigan State University

James M. Larson
Michigan State University

Ronald Preston
Indiana University

Tat Ming Sze
Michigan State University

Sarah Theule-Lubienski
Michigan State University

Jeffrey J. Wanko
Michigan State University

Field Test Production Team

Katherine Oesterle
Michigan State University

Stacey L. Otto
University of North Carolina at Chapel Hill

Teacher/Assessment Team

Kathy Booth
Waverly, Michigan

Anita Clark
Marshall, Michigan

Theodore Gardella
Bloomfield Hills, Michigan

Yvonne Grant
Portland, Michigan

Linda R. Lobue
Vista, California

Suzanne McGrath
Chula Vista, California

Nancy McIntyre
Troy, Michigan

Linda Walker
Tallahassee, Florida

Software

Richard Bu
East Lansing, Michigan

Development Center Directors

Nicholas Branca
San Diego State University

Dianne Briars
Pittsburgh Public Schools

Frances R. Curcio
New York University

Perry Lanier
Michigan State University

J. Michael Shaughnessy
Portland State University

Charles Vonder Embse
Central Michigan University

Field Test Coordinators

Michelle Bohan
Queens, New York

Melanie Branca
San Diego, California

Alecia Devantier
Shepherd, Michigan

Jenny Jorgensen
Flint, Michigan

Sandra Kralovec
Portland, Oregon

Sonia Marsalis
Flint, Michigan

William Schaeffer
Pittsburgh, Pennsylvania

Karma Vince
Toledo, Ohio

Virginia Wolf
Pittsburgh, Pennsylvania

Shirel Yaloz
Queens, New York

Student Assistants

Laura Hammond
David Roche
Courtney Stoner
Jovan Trpovski
Julie Valicenti
Michigan State University

Greg Williams
Gundry Elementary School

Lansing

Susan Bissonette
Waverly Middle School

Kathy Booth
Waverly East Intermediate School

Carole Campbell
Waverly East Intermediate School

Gary Gillespie
Waverly East Intermediate School

Denise Kehren
Waverly Middle School

Virginia Larson
Waverly East Intermediate School

Kelly Martin
Waverly Middle School

Laurie Metevier
Waverly East Intermediate School

Craig Paksi
Waverly East Intermediate School

Tony Pecoraro
Waverly Middle School

Helene Rewa
Waverly East Intermediate School

Arnold Stiefel
Waverly Middle School

Portland

Bill Carlton
Portland Middle School

Kathy Dole
Portland Middle School

Debby Flate
Portland Middle School

Yvonne Grant
Portland Middle School

Terry Keusch
Portland Middle School

John Manzini
Portland Middle School

Mary Parker
Portland Middle School

Scott Sandborn
Portland Middle School

Shepherd

Steve Brant
Shepherd Middle School

Mary Brock
Shepherd Middle School

Cathy Church
Shepherd Middle School

Ginny Crandall
Shepherd Middle School

Craig Ericksen
Shepherd Middle School

Natalie Hackney
Shepherd Middle School

Bill Hamilton
Shepherd Middle School

Julie Salisbury
Shepherd Middle School

Sturgis

Sandra Allen
Eastwood Elementary School

Margaret Baker
Eastwood Elementary School

Steven Baker
Eastwood Elementary School

Keith Barnes
Eastwood Elementary School

Wilodean Beckwith
Eastwood Elementary School

Darcy Bird
Eastwood Elementary School

Bill Dickey
Sturgis Middle School

Ellen Eisele
Eastwood Elementary School

James Hoelscher
Sturgis Middle School

Richard Nolan
Sturgis Middle School

J. Hunter Raiford
Sturgis Middle School

Cindy Sprowl
Eastwood Elementary School

Leslie Stewart
Eastwood Elementary School

Connie Sutton
Eastwood Elementary School

Traverse City

Maureen Bauer
Interlochen Elementary School

Ivanka Berskshire
East Junior High School

Sarah Boehm
Courtade Elementary School

Marilyn Conklin
Interlochen Elementary School

Nancy Crandall
Blair Elementary School

Fran Cullen
Courtade Elementary School

Eric Dreier
Old Mission Elementary School

Lisa Dzierwa
Cherry Knoll Elementary School

Ray Fouch
West Junior High School

Ed Hargis
Willow Hill Elementary School

Richard Henry
West Junior High School

Dessie Hughes
Cherry Knoll Elementary School

Ruthanne Kladder
Oak Park Elementary School

Bonnie Knapp
West Junior High School

Sue Laisure
Sabin Elementary School

Stan Malaski
Oak Park Elementary School

Jody Meyers
Sabin Elementary School

Marsha Myles
East Junior High School

Mary Beth O'Neil
Traverse Heights Elementary School

Jan Palkowski
East Junior High School

Karen Richardson
Old Mission Elementary School

Kristin Sak
Bertha Vos Elementary School

Mary Beth Schmitt
East Junior High School

Mike Schrotenboer
Norris Elementary School

Gail Smith
Willow Hill Elementary School

Karrie Tufts
Eastern Elementary School

Mike Wilson
East Junior High School

Tom Wilson
West Junior High School

Minnesota

Minneapolis

Betsy Ford
Northeast Middle School

New York

East Elmhurst

Allison Clark
Louis Armstrong Middle School

Dorothy Hershey
Louis Armstrong Middle School

J. Lewis McNeece
Louis Armstrong Middle School

Rossana Perez
Louis Armstrong Middle School

Merna Porter
Louis Armstrong Middle School

Marie Turini
Louis Armstrong Middle School

North Carolina

Durham

Everly Broadway
Durham Public Schools

Thomas Carson
Duke School for Children

Mary Hebrank
Duke School for Children

Bill O'Connor
Duke School for Children

Ruth Pershing
Duke School for Children

Peter Reichert
Duke School for Children

Elizabeth City

Rita Banks
Elizabeth City Middle School

Beth Chaundry
Elizabeth City Middle School

Amy Cuthbertson
Elizabeth City Middle School

Deni Dennison
Elizabeth City Middle School

Jean Gray
Elizabeth City Middle School

John McMenamin
Elizabeth City Middle School

Nicollette Nixon
Elizabeth City Middle School

Malinda Norfleet
Elizabeth City Middle School

Joyce O'Neal
Elizabeth City Middle School

Clevie Sawyer
Elizabeth City Middle School

Juanita Shannon
Elizabeth City Middle School

Terry Thorne
Elizabeth City Middle School

Rebecca Wardour
Elizabeth City Middle School

Leora Winslow
Elizabeth City Middle School

Franklinton

Susan Haywood
Franklinton Elementary School

Clyde Melton
Franklinton Elementary School

Louisburg

Lisa Anderson
Terrell Lane Middle School

Jackie Frazier
Terrell Lane Middle School

Pam Harris
Terrell Lane Middle School

Ohio

Toledo

Bonnie Bias
Hawkins Elementary School

Marsha Jackish
Hawkins Elementary School

Lee Jagodzinski
DeVeaux Junior High School

Norma J. King
Old Orchard Elementary School

Margaret McCready
Old Orchard Elementary School

Carmella Morton
DeVeaux Junior High School

Karen C. Rohrs
Hawkins Elementary School

Marie Sahloff
DeVeaux Junior High School

L. Michael Vince
McTigue Junior High School

Brenda D. Watkins
Old Orchard Elementary School

Oregon

Portland

Roberta Cohen
Catlin Gabel School

David Ellenberg
Catlin Gabel School

Sara Normington
Catlin Gabel School

Karen Scholte-Arce
Catlin Gabel School

West Linn

Marge Burack
Wood Middle School

Tracy Wygant
Athey Creek Middle School

Canby

Sandra Kralovec
Ackerman Middle School

Pennsylvania

Pittsburgh

Sheryl Adams
Reizenstein Middle School

Sue Barie
Frick International Studies Academy

Suzie Berry
Frick International Studies Academy

Richard Delgrosso
Frick International Studies Academy

Janet Falkowski
Frick International Studies Academy

Joanne George
Reizenstein Middle School

Harriet Hopper
Reizenstein Middle School

Chuck Jessen
Reizenstein Middle School

Ken Labuskes
Reizenstein Middle School

Barbara Lewis
Reizenstein Middle School

Sharon Mihalich
Reizenstein Middle School

Marianne O'Conner
Frick International Studies Academy

Mark Sammartino
Reizenstein Middle School

Washington

Seattle

Chris Johnson

University Preparatory Academy

Rick Purn
University Preparatory Academy

Contents

Rational numbers are at the heart of the middle-grades experience with number concepts. The concepts of fractions, decimals, and percents are often difficult for students. Research tells us that part of the reason for students' confusion about rational numbers is a consequence of the rush to symbol manipulation with fractions and decimals. Students need time to develop a deep understanding of fractions and decimals. The investigations in *Bits and Pieces I* ask students to make sense of fractions, decimals, and percents in different contexts.

The many different and powerful interpretations of and models for rational numbers can make grasping ideas about such numbers difficult. To gain a mature knowledge of rational numbers, students must be able to handle these various interpretations. We have carefully chosen the interpretations and models used in the unit. Some models are more powerful than others, as they contribute to developing the meaning of rational numbers and to understanding operations on rational numbers.

This unit does not teach specific algorithms for work with rational numbers. Instead, it helps the teacher create a supportive environment for students to grapple with interesting problems in which ideas of fractions, decimals, and percents are imbedded. As students work—individually, in groups, and as a class—they will develop ways of thinking about rational numbers. The teacher's role is to help students make explicit their growing ideas about this world of rational numbers. The intent of this unit is to provide a rich set of experiences that focus on developing meaning.

In this unit, students will meet several interpretations and models of fractions. These have been carefully chosen so that the move between problems will add to a deepening knowledge and comfort with fractions.

Interpretations of Fractions

The major interpretations on which this unit focuses are

- fractions as parts of a whole

- fractions as measures or quantities

- fractions as indicated division

- fractions as decimals

- fractions as percents

Other interpretations—such as fractions as operators ("stretchers" or "shrinkers") and fractions as rates, ratios, or parts of a proportion—are postponed until later grades.

Fractions as Parts of a Whole

This interpretation of rational numbers is applied in situations that are continuous and in situations that consider discrete objects. The important characteristic is that this interpretation depends on partitioning an object or a set into equal-size parts and making a comparison of some of the parts to the whole object or set. For example, if there are 27 students in the class and 13 are girls, the part of the whole that is girls can be represented as $\frac{13}{27}$.

In the following diagram, two parts are shaded.

The shaded portion can be represented as $\frac{2}{3}$. The 3 tells into how many equal-size parts the whole has been divided, and the 2 tells how many of the equal-size parts have been shaded.

In the part-whole interpretation of fractions, the difficulties for students center on the following:

- determining what the whole is

- subdividing the whole into equal-size parts—not equal *shape*, but equal *size*

- recognizing how many parts are needed to represent the situation

- forming the fraction by placing the parts needed over the number of parts into which the whole has been divided

Fractions as Measures or Quantities

In this interpretation, a fraction is thought of as a number. For example, a fraction can be a measurement that is "in between" two whole measures. Students meet this every day in such references as $2\frac{1}{2}$ feet or 11.5 million people. Understanding this interpretation is important for students' mathematical development, and it leads to comparison of fractions and operations on fractions.

Fractions as Indicated Divisions

To move with flexibility between fraction and decimal representations of rational numbers, students need to understand how fractions can be thought of as indicated divisions. Sharing is a natural context in which to help students see how this interpretation is related to whole-number division. If students see that sharing 36 apples among 6 people calls for division ($36 \div 6 = 6$ apples each), then they can move to an understanding that sharing 3 apples among 8 people calls for dividing 3 by 8 to find out how many each person receives.

Fractions as Decimals

A byproduct of the division interpretation of fractions is the relationship between a fraction and decimal representation of the same quantity. For the fraction $\frac{2}{5}$, for example, we can find the decimal representation by dividing 2 by 5. Given the modern tools of calculators and computers, decimal representations are even more important today than in the past. Students need time to develop comfort and ease in moving between fractions and decimals, and they need to understand decimals in two ways:

- as special fractions with denominators of 10 and powers of 10

- as a natural extension of the place-value system for representing quantities less than 1

Fractions as Percents

Rather than treating fractions, decimals, and percents as separate topics, this unit seeks to build the connections between them. Students will see that the ideas and concepts are related and that the differences are in the symbols used to represent those ideas. Ten percent, 10%, is simply another way to represent 0.10 or 0.1, which is another way to represent $\frac{10}{100}$ or $\frac{1}{10}$. Percents are introduced as special names for parts of 100.

Models of Fractions

The models of rational numbers used throughout this unit were chosen because they connect directly to the interpretations of rational numbers that the unit raises. The models on which this unit focuses are

- fractions-strip models

- number-line models

- grid-area models

- partition models

Fraction-Strip Models

Students are introduced to fractions in a situation that uses a *fraction strip* as a model. Fraction strips can be created by dividing a strip of paper into equal-size parts by folding. This is a fraction strip for halves:

$$\frac{1}{2}$$

Number-Line Models

The collection of fraction strips are used to move to a number-line model of rational numbers. The *number-line model* helps make the connection to fractions as numbers or quantities. This is a number line for 0 to 2 with a few fractional quantities marked:

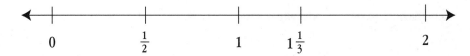

Grid-Area Models

Because 100 and powers of 10 are so useful in understanding decimals and percents, *grid-area models* are introduced and developed in this unit. This grid shows a shaded area of 12%.

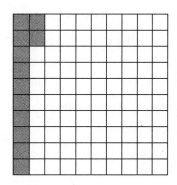

Partition Models

Students also use a more general model of fraction situations that is based on *partitioning an area*, such as a circle, into equal-size parts. The circle shows a shaded portion of $\frac{3}{10}$.

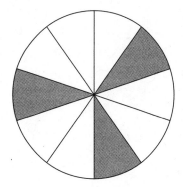

Bits *and Pieces I* **was created to help students**

- Build an understanding of fractions, decimals, and percents and the relationships between and among these concepts and their representations

- Develop ways to model situations involving fractions, decimals, and percents

- Understand and use equivalent fractions to reason about situations

- Compare and order fractions

- Move flexibly between fraction, decimal, and percent representations

- Use 0, $\frac{1}{2}$, 1, and $1\frac{1}{2}$ as benchmarks to help estimate the size of a number or sum

- Develop and use benchmarks that relate different forms of representations of rational **numbers** (for example, 50% is the same as $\frac{1}{2}$ and 0.5)

- Use physical models and drawings to help reason about a situation

- Look for patterns and describe how to continue the pattern

- Use context to help reason about a situation

- Use estimation to understand a situation

Investigation 1: Fund-Raising Fractions

Students explore three components of understanding fractions: the visual model (fraction strips), word names for fractions, and symbols for fractions. The part-whole interpretation of fractions is developed. Students make fraction strips to study the progress toward a fund-raising goal. The aim is to focus on the meaning of such phrases as, "two thirds of the goal has been reached."

Investigation 2: Comparing Fractions

The most important concept in understanding and using rational numbers is equivalence of fractions. This concept underlies operations with fractions, changing representations of fractions, and reasoning proportionally. The context of comparing fraction strips is used to motivate an investigation of equivalence and the creation of a number line that contains all of the information of the individual fraction strips. The idea of using benchmarks to estimate the size of fractions and to make comparisons is introduced.

Investigation 3: Cooking with Fractions

The context of cooking—parts of cups or other measures often called for in recipes, and the need to make multiples of a recipe, sets the stage for introducing students to different kinds of area models for fractions. The square and the rectangle are particularly useful areas because they are easy to subdivide and to shade. The circle is explored because of its use in data analysis and probability.

Investigation 4: From Fractions to Decimals

Students are introduced to decimal representations of fractions and explore the place-value interpretation of decimals. They investigate a 100-square grid and explore how it could continue to be subdivided to show 1000 parts or 10,000 parts. This process of subdividing and naming the new parts is very important mathematically; the underpinnings of the infinite process are met in this problem. The process will continue to help students understand equivalence of fraction and equivalence of decimals as well as to see the connections between fractions and decimals.

Investigation 5: Moving Between Fractions and Decimals

This investigation proposes a situation in which fractions with denominators larger than students' fraction strips show must be compared. Students find decimal estimates for fractions using the visual model. They are asked to consider whether fractions or decimals are easier to compare. Sharing is used as a context to motivate the division interpretation of fractions, leading to a strategy for changing a fraction into a decimal. Calculators are used to do the computation, providing additional evidence that the division interpretation as a way to find decimal equivalents makes sense.

Investigation 6: Out of One Hundred

By this time, students should feel comfortable with the meaning of fractions and decimals and be able to move back and forth between the two. Percents are now introduced as another form of representation. A database of information about cats is used as a context for understanding percent. Students are engaged in activities requiring them to move among fractions, decimals, and percents.

Materials

For students

- Labsheets
- $8\frac{1}{2}$" strips of paper for making fraction strips
- Distinguishing Digits puzzle cards (provided as blackline masters)
- Scissors
- Rulers or other straightedges
- Colored cubes or tiles (optional)
- Blank transparency film (optional)
- Large sheets of blank paper (optional)
- Index cards (optional)
- Grid paper (optional; provided as a blackline master)

For the teacher

- Transparencies and transparency markers (optional)
- $8\frac{1}{2}$" fraction strips for the overhead projector
- 16 cm fraction strips for the overhead projector (optional; copy Labsheet 1.5 onto blank transparency film)
- $5\frac{2}{3}$" strips of paper (optional)
- A transparent centimeter ruler (optional)
- Transparency of newspaper advertisement (optional)

Technology

The Connected Mathematics Project was developed with the belief that calculators should always be available and that students should decide when to use them. For this reason, we do not designate specific problems as "calculator problems." Fraction calculators are *not* required. However, if fraction calculators are available, your students can use them as an additional tool for exploring the ideas of this unit.

Pacing Chart

This pacing chart gives estimates of the class time required for each investigation and assessment piece. Shaded rows indicate opportunities for assessment.

Investigations and Assessments	Class Time
1 Fund-Raising Fractions	5 days
2 Comparing Fractions	5 days
Check-Up 1	$\frac{1}{2}$ day
3 Cooking with Fractions	2 days
Quiz	1 day
4 From Fractions to Decimals	3 days
5 Moving Between Fractions and Decimals	4 days
Check-Up 2	$\frac{1}{2}$ day
6 Out of One Hundred	4 days
Self-Assessment	Take home
Unit Test	1 day

Bits and Pieces I Vocabulary

The following words and concepts are introduced and used in *Bits and Pieces I*. Concepts in the left column are those that are essential for student understanding of this and future units. The Descriptive Glossary/Index gives descriptions of these and other words used in *Bits and Pieces I*.

Essential	Nonessential
decimal	base ten number system
denominator	benchmark
equivalent fraction	unit fraction
fraction	
numerator	
percent	

Embedded Assessment

Opportunities for informal assessment of student progress are embedded throughout *Bits and Pieces I* in the problems, the ACE questions, and the Mathematical Reflections. Suggestions for observing as students discover and explore mathematical ideas, for probing to guide their progress in developing concepts and skills, and for questioning to determine their level of understanding can be found in the *Launch, Explore,* or *Summarize* sections of all investigation problems. Some examples:

- Investigation 4, Problem 4.2 *Launch* (page 52b) suggests ways to assess your students's understanding of decimals.
- Investigation 2, Problem 2.5 *Explore* (page 30h) suggests questions you might ask to help your students think about how to label points on the number line with improper fractions.
- Investigation 1, Problem 1.4 *Summarize* (page 18h) suggests questions you might ask to assess your students' understanding of fractions of different wholes.

ACE Assignments

An ACE (Applications—Connections—Extensions) section appears at the end of each investigation. To help you assign ACE questions, a list of assignment choices is given in the margin next to the reduced student page for each problem. Each list indicates the ACE questions that students should be able to answer after they complete the problem.

Partner Quiz

One quiz, which may be given after Investigation 3, is provided with *Bits and Pieces I*. This quiz is designed to be completed by pairs of students with the opportunity for revision based on teacher feedback. You will find the quiz and its answer key in the Assessment Resources section. As an alternative to the quiz provided, you can construct your own quizzes by combining questions from the Question Bank, the quiz, and unassigned ACE questions.

Check-Ups

Two check-ups, which may be given after Investigations 2 and 5, are provided for use as quick quizzes or warm-up activities. Check-ups are designed for students to complete individually. You will find the check-ups, their answer keys, and a guide for assessing the check-ups in the Assessment Resources section.

Question Bank

A Question Bank provides questions you can use for homework, reviews, or quizzes. You will find the Question Bank and its answer key in the Assessment Resources section.

Notebook/Journal

Students should have notebooks to record and organize their work. In the notebooks will be their journals along with sections for vocabulary, homework, and quizzes and check-ups. In their journals, students can take notes, solve investigation problems, and record their mathematical reflections. You should assess student journals for completeness rather than correctness; journals should be seen as "safe" places where students can try out their thinking. A Notebook Checklist and a Self-Assessment are provided in the Assessment Resources section. The Notebook Checklist helps students organize their notebooks. The Self-Assessment guides students as they review their notebooks to determine which ideas they have mastered and which ideas they still need to work on.

The Unit Test

The final assessment for *Bits and Pieces I* is a two-part test. The first part is an individual, in-class unit test. The second part is a short individual research and writing assignment and is meant to be treated as a take-home portion to the test.

For the research and writing piece, students are to find two different articles that contain fractions, decimals, and/or percents. They are to write a one- to two-paragraph summary of each article, explaining how the rational numbers are used in the articles and what they represent.

Blackline masters for the in-class and take-home portions of the unit test, as well as an answer key for the in-class test, are provided in the Assessment Resources section.

Introducing Your Students to *Bits and Pieces I*

Several days before starting this unit, ask your students to find examples of how fractions, decimals, or percents are used in everyday life. Encourage them to look at advertisements, magazines, or newspapers or to interview an adult. On the day you start the unit, let your students present the examples they found. Ask students to explain what they think the fractions, decimals, or percents in their examples mean. Don't worry about getting correct explanations at this time; this excercise is meant to get students interested in this unit and to start them thinking about fractions. If you think an example is especially interesting, you may want to return to it later in the unit, when students can apply what they have learned to make sense of it.

When everyone has had a chance to share his or her example, discuss the three questions posed on the opening pages of the student edition. It is not necessary to come up with correct answers at this time. These questions are raised later in the unit when students have the skills necessary to answer them.

Bits and Pieces I

In a survey of 200 cat owners, 59% said they fed their pet only pet food, not human food. How many people is this?

Samuel is getting a snack for himself and his little brother. There are two candy bars in the refrigerator. Samuel takes half of one candy bar for himself and half of the other candy bar for his little brother. His little brother complains that Samuel got more. Samuel says that he got half and his brother got half. What might be the problem?

Tisia made 19 out of 30 free throws; Clarise made 8 out of 13; and Dorothea made 14 out of 21. Who is the best free-throw shooter?

Tips for the Linguistically Diverse Classroom

Enactment The Enactment technique is described in detail in *Getting to Know CMP*. Students act out mini-scenes, using props, to make information comprehensible. Example: For the question about the pet food survey, one student could survey other students about what type of food they feed their pets. A survey, a can of pet food, and a sample of human food could be used as props.

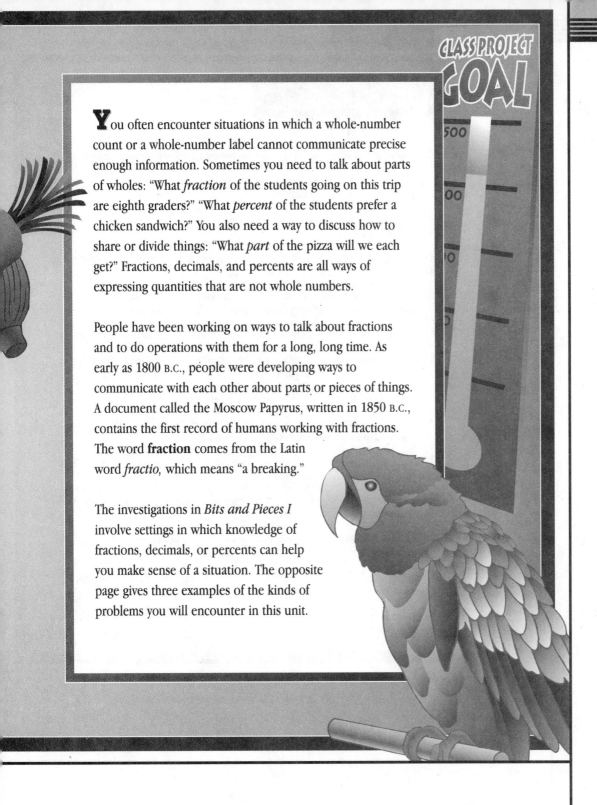

You often encounter situations in which a whole-number count or a whole-number label cannot communicate precise enough information. Sometimes you need to talk about parts of wholes: "What *fraction* of the students going on this trip are eighth graders?" "What *percent* of the students prefer a chicken sandwich?" You also need a way to discuss how to share or divide things: "What *part* of the pizza will we each get?" Fractions, decimals, and percents are all ways of expressing quantities that are not whole numbers.

People have been working on ways to talk about fractions and to do operations with them for a long, long time. As early as 1800 B.C., people were developing ways to communicate with each other about parts or pieces of things. A document called the Moscow Papyrus, written in 1850 B.C., contains the first record of humans working with fractions. The word **fraction** comes from the Latin word *fractio,* which means "a breaking."

The investigations in *Bits and Pieces I* involve settings in which knowledge of fractions, decimals, or percents can help you make sense of a situation. The opposite page gives three examples of the kinds of problems you will encounter in this unit.

Mathematical Highlights

The Mathematical Highlights page provides information to students and to parents and other family members. It gives students a preview of the activities and problems in *Bits and Pieces I*. As they work through the unit, students can refer back to the Mathematical Highlights page to review what they have learned and to preview what is still to come. This page also tells parents and other family members what mathematical ideas and activities will be covered as the class works through *Bits and Pieces I*.

Mathematical Highlights

In *Bits and Pieces I*, you will learn to represent and talk about fractions.

- Interpreting a thermometer display, which shows the progress of a sixth-grade fund-raising campaign, helps you understand fractions as parts of wholes.

- Fraction strips, which you fold and label yourself, give you a concrete model for visualizing fractions.

- Comparing fund-raisers with different goals demonstrates the importance of knowing the "whole" when comparing fractions.

- Transferring the fractions from all of your strips onto a single number line lets you compare fractions with different denominators and find equivalent fractions.

- Using common fractions as benchmarks helps you to compare and estimate fractions.

- Finding the possible ways to cut a pan of brownies into equal pieces shows you that fractions can be thought of as parts of regions.

- Planning a vegetable garden helps you think about expressing fractions as decimals.

- What you know about decimals, place value, factors, and multiples enables you to solve some challenging Distinguishing Digits Puzzles.

- Finding equivalent fractions allows you to choose the best shooter to attempt a game-winning free throw.

- Using a new fraction strip, the hundredths strip, helps you find decimal estimates for all the marks on your other fraction strips.

- Distributing food items among care packages leads you to discover that a fraction is a way to indicate a division.

- A database of information about 100 cats introduces you to percents as a way to represent fractions.

- Rewriting sale signs for a pet store lets you apply what you have learned about moving between fractions, decimals, and percents.

The Investigations

The teaching materials for each investigation consist of three parts: an overview, the student pages with teaching outlines, and the detailed notes for teaching the investigation.

The overview of each investigation includes brief descriptions of the problems, the mathematical and problem-solving goals of the investigation, and a list of necessary materials.

Essential information for teaching the investigation is provided in the margins around the student pages. The "At a Glance" overviews are brief outlines of the Launch, Explore, and Summarize phases of each problem for reference as you work with the class. To help you assign homework, a list of "Assignment Choices" is provided next to each problem. Wherever space permits, answers to problems, follow-ups, ACE questions, and Mathematical Reflections appear next to the appropriate student pages.

The Teaching the Investigation section follows the student pages and is the heart of the Connected Mathematics curriculum. This section describes in detail the Launch, Explore, and Summarize phases for each problem. It includes all the information needed for teaching, along with suggestions for what you might say at key points in the teaching. Use this section to prepare lessons and as a guide for teaching an investigation.

Assessment Resources

The Assessment Resources section contains blackline masters and answer keys for the quiz, check-ups, the Question Bank, and the Unit Test. Blackline masters for the Notebook Checklist and the Self-Assessment are also given. These instruments support student self-evaluation, an important aspect of assessment in the Connected Mathematics curriculum. This section also includes a guide to assessing the check-ups and a discussion about altering assessment pieces for inclusion students.

Blackline Masters

The Blackline Masters section includes masters for all labsheets and transparencies. Blackline masters of grid paper and Distinguishing Digits puzzle cards are also provided.

Descriptive Glossary/Index

The Descriptive Glossary/Index provides descriptions and examples of the key concepts in *Bits and Pieces I.* These descriptions are not intended to be formal definitions, but are meant to give you an idea of how students might make sense of these important concepts. The page number references indicate where each concept is first introduced.

Fund-Raising Fractions

Mathematical and Problem-Solving Goals

- **To use the part-whole interpretation of fractions to create a set of fraction strips**

- **To relate the fraction-strip model to the part-whole interpretation of fractions and to the symbolic representation of fractions**

- **To understand the meaning of fractions larger than a whole**

- **To use fraction strips and symbolic representations of fractions to describe real-world situations**

In this lengthy but important investigation, students use fractions to describe the progress of various fund-raising activities at a typical school. This investigation develops the part-whole interpretation of fractions, explores the fraction-strip model for representing fractions, and shows how the symbolic representation of fractions relates to a physical model (the fraction strips).

In Problem 1.1, Reporting Our Progress, students write short reports describing the progress the sixth graders at Thurgood Marshall School have made toward their fund-raising goal. This activity lets you quickly assess your students' understanding of fractions as parts of wholes. In Problem 1.2, Using Fraction Strips, students are challenged to make fraction strips by folding paper and then to use these strips to investigate the progress of the sixth-grade fund-raiser at various stages. In Problem 1.3, Comparing Classes, students explore comparing fractions with different wholes. Problem 1.4, Exceeding the Goal, involves a fundraiser in which the amount of money raised surpassed the goal; students must describe situations involving fractions greater than 1. Problem 1.5, Using Symbolic Form, begins to develop the number-line model of fractions. Students label parts of their fraction strips and begin to think about the meaning of the symbolic representation of fractions.

Student Pages	5–18
Teaching the Investigation	18a–18k

Materials

For students

- Labsheet 1.5 (1 per student)
- $8\frac{1}{2}$" strips of paper (9 per student)

For the teacher

- Transparencies 1.1, 1.2A, 1.2B, 1.3A, 1.3B, 1.4A, 1.4B, and 1.5 (optional)
- $8\frac{1}{2}$" fraction strips for the overhead projector
- Transparent centimeter ruler (optional)
- 16 cm fraction strips for the overhead projector (optional; copy Labsheet 1.5 onto blank transparency film)

INVESTIGATION 1

Goal — $300

Fund-Raising Fractions

Last year students at Thurgood Marshall School organized three fund-raising projects to raise money for sports and band equipment. The eighth-grade class held a calendar sale in October, the seventh-grade class sold popcorn in January, and the sixth-grade class sold art, music, and sports posters in March. The three grades competed to raise the most money.

1.1 Reporting Our Progress

The sixth-grade class set a goal of raising $300 during its ten-day poster sale. On each day of the sale, the class's progress was marked on a large "thermometer" near the school office.

The thermometer at right shows the progress of the sixth-grade fund-raiser after two days of sales. The goal of $300 is marked near the top of the thermometer. Every day during the fund-raising campaign, the sixth-grade class officers gave a public-address announcement reporting their progress.

Problem 1.1

Write a short—but clever and informative—announcement to report the progress of the sixth-grade poster sale after two days. Be sure to mention what part of the sales goal of $300 had been reached and what part remained to be raised.

▥ Problem 1.1 Follow-Up

Describe the strategies you used to decide what part of the sales goal of $300 had been reached and what part remained to be raised.

Day 2

Sixth-Grade
Poster Sale

■■■ At a Glance

Launch

■ Discuss the thermometer model often used to track the progress of fund-raisers.

■ Introduce the story problem presented in the student edition.

Explore

■ Circulate while students write public-address announcements, helping them to clarify their messages.

Summarize

■ Have a few groups present and explain their announcements.

■ As a class, analyze the strategies and information in the different announcements.

Answer to Problem 1.1

Answers will vary depending on how comfortable your students are with fractions, decimals, and percents. Your students may not be able to respond in a precise mathematical way. The thermometer is registering 25%, or one fourth ($\frac{1}{4}$), of the goal, so students have raised $75 of the $300 needed. The students still need to meet 75%, or three fourths ($\frac{3}{4}$), of their goal; they must still raise $225 of the $300. Some students may comment on the fact that the sixth-grade class has raised one fourth ($\frac{1}{4}$) of the goal in only one fifth ($\frac{1}{5}$) of the time allotted for the campaign.

Answers to Problem 1.1 Follow-Up

See page 18j.

Assignment Choices

Challenge students to look for examples of fractions and decimals used outside of school.

Using Fraction Strips

At a Glance

Launch

- Discuss the thermometers showing the progress of the sixth-grade fund-raisers.

- Demonstrate how to make fraction strips for halves and thirds.

Explore

- Circulate while students explore techniques for making fraction strips.

- As a class, discuss the various folding techniques students found, focusing on the concepts of part and whole.

- Have students continue their work on the problem.

Summarize

- Have students share their estimates and strategies.

- Focus students on how fraction strips allow us to estimate parts of wholes.

The thermometers on the next page show the progress of the sixth-grade poster sale after two, four, six, eight, and ten days. One way to determine the progress of the fund-raiser is to use strips of paper the same length as the distance from the bottom of the thermometer to the goal. By folding the strips into fractional parts, you can determine what part of the goal has been reached.

Problem 1.2

Start with nine $8\frac{1}{2}$-inch strips. Fold the strips to show halves, thirds, fourths, fifths, sixths, eighths, ninths, tenths, and twelfths. Mark the folds in the strips with a pencil so you can see them more easily.

Use your strips to estimate the sixth-grade class's progress after two, four, six, eight, and ten days.

■ **Problem 1.2 Follow-Up**

1. Which fraction strips were easy to fold? Why?
2. Which fraction strips were difficult to fold? Why?

Assignment Choices

ACE questions 1 and 26

Answer to Problem 1.2

On day 2, the students had reached one fourth, two eighths, or three twelfths of their goal. On day 4, they had reached one third, two sixths, three ninths, or four twelfths of their goal. On day 6, they had reached three fifths or six tenths of their goal. On day 8, they had reached three fourths, six eighths, or nine twelfths of their goal. On day 10, they had reached five sixths or ten twelfths of their goal.

Answers to Problem 1.2 Follow-Up

1. Answers will vary. Students may indicate that the halves, fourths, and eighths strips were easy to fold because they simply required repeatedly folding in half.

2. Answers will vary. Students may indicate that the thirds, fifths, and ninths strips (and perhaps the sixths, tenths, and twelfths strips) were hard to fold because they could not be made by simply folding strips in half.

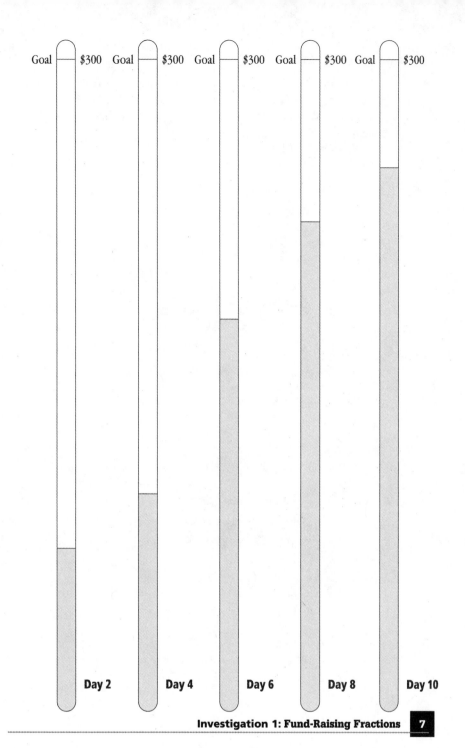

Investigation 1: Fund-Raising Fractions 7

1.3

Comparing Classes

At a Glance

Launch

- Talk about the thermometers showing the results of the three grades' fund-raisers.

- Discuss how to determine from the thermometer the amount of money the sixth graders raised.

Explore

- Circulate while students explore strategies for evaluating the seventh and eighth graders' claims.

- Look for students who are using effective strategies and for students who are having difficulty.

Summarize

- As a class, compare the various arguments proposed by the groups.

- Have groups present arguments for each grade level.

- Ask students to predict how long the fund-raisers would need to continue for each grade to reach its goal.

In Thurgood Marshall School, the seventh-grade class is larger than the sixth-grade class, and the eighth-grade class is smaller than the sixth-grade class. Because they are different sizes, each class set a different goal for its fund-raiser. The sixth grade set a goal of $300 for its poster sale, the seventh grade set a goal of $400 for its popcorn sale, and the eighth grade set a goal of $240 for its calendar sale.

The thermometers on the next page show the results of the sixth-grade, seventh-grade, and eighth-grade fund-raisers. Both the seventh graders and the eighth graders claimed to do better than the sixth graders.

Problem 1.3

Use the fraction strips you made in Problem 1.2 to investigate the seventh and eighth graders' claims.

A. How much money did each grade raise?

B. What fraction of the goal did each grade reach?

C. What argument could the eighth graders use to claim that their class did better than the sixth grade?

D. What argument could the seventh graders use to claim that their class did better than the sixth grade?

■ Problem 1.3 Follow-Up

Which of the three classes do you think did the best job? Explain your reasoning.

Assignment Choices

ACE questions 13–25 and unassigned choices from earlier problems

Answers to Problem 1.3

A. The seventh-grade class raised about $300. The eighth-grade class raised about $220.

B. The seventh-grade class is about three fourths ($\frac{3}{4}$) of the way to its goal. The eighth-grade class is about eleven twelfths ($\frac{11}{12}$) of the way to its goal.

C. The eighth graders can say they are closer to their goal.

D. The seventh graders can say they have raised more money.

Answer to Problem 1.3 Follow-Up

Answers will vary. Some students may say that the eighth grade is closer to their goal even though they have raised less money than the other grades. Other students may pick the seventh grade because, although they have reached a smaller fraction of their goal, they have raised more money.

Day 10

Sixth-Grade
Poster Sale

Day 10

Seventh-Grade
Popcorn Sale

Day 10

Eighth-Grade
Calendar Sale

Investigation 1: Fund-Raising Fractions 9

1.4

Exceeding
the Goal

At a Glance

Launch

■ As a class, examine the teachers' thermometers, focusing on the facts that the teachers' thermometer is shorter than the students' and that the teachers exceeded their goal.

Explore

■ Circulate while students explore strategies for analyzing the teachers' thermometers.

■ Note any creative strategies students use.

Summarize

■ Have students share their strategies for working with the new whole.

■ Ask questions to extend the class's thinking about the part-whole meaning of fractions.

Assignment Choices

ACE questions 8–11 and unassigned choices from earlier problems

1.4) **Exceeding the Goal**

In April, the Thurgood Marshall School needed money to put on the annual Year's End Festival. Since the students had worked hard on the earlier fund-raising campaigns, the teachers volunteered to raise the money. They decided to sell paperback books for summer reading, and they set a goal of $360.

The thermometers on the next page show the teachers' progress at the end of the second, sixth, and tenth days.

Problem 1.4

A. Notice that the teachers used a shorter thermometer than the students did to report their progress. Can you use your fraction strips to measure these thermometers? Explain.

B. What fraction of their goal did the teachers reach at the end of each of the days shown? Explain how you determined your answers.

C. How many dollars did the teachers raise by the end of each of these days?

■ Problem 1.4 Follow-Up
What school announcement might the teachers make at the end of the tenth day?

Bits and Pieces I

Answers to Problem 1.4

A. Since the distance to the goal for the teachers' thermometer is not the length of the fraction strips used to measure the students' thermometers, these strips cannot be used *in the same way* to reason about the teachers' thermometers. Some students may come up with other ways to use these strips (see the discussion in the "Explore" section on page 18h).

B. At the end of day 2, the teachers were one fourth ($\frac{1}{4}$) of the way to their goal. At the end of day 6, the teachers had reached their goal. At the end of day 10, the teachers had reached five fourths ($\frac{5}{4}$), or one and one fourth ($1\frac{1}{4}$), of their goal. Explanations of strategies will vary.

C. Teachers raised $90 at the end of day 2, $360 at the end of day 6, and $450 at the end of day 10.

Investigation 1

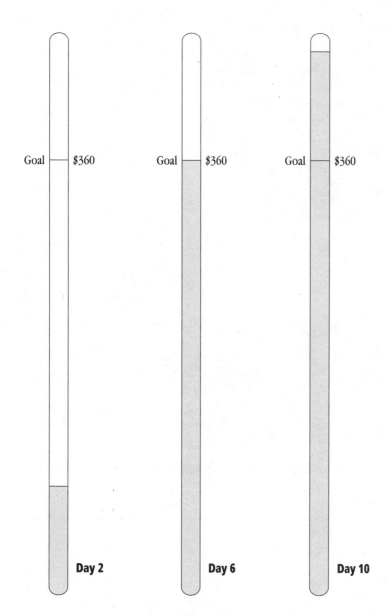

Day 2

Day 6

Day 10

Teachers' Book Sale

Answer to Problem 1.4 Follow-Up

Answers will vary. The announcement should mention that the teachers exceeded their goal; they have raised about $450, which is $1\frac{1}{4}$ of their goal.

Using Symbolic Form

Fractions can be written in *symbolic form,* using two whole numbers separated by a bar. For example, one half is written $\frac{1}{2}$ and two thirds is written $\frac{2}{3}$.

The number above the bar is called the **numerator,** and the number below the bar is called the **denominator.**

At a Glance

Launch

- Discuss the accuracy of the students' fraction strips.

- Talk about the meaning of the numerator and the denominator in a fraction.

Explore

- Circulate while pairs label the fraction strips on Labsheet 1.5.

- Ask questions to focus students on the meaning of the symbols they are writing.

Summarize

- As a class, compare the printed fraction strips with the strips the students folded and with a centimeter ruler.

Think about this!

What do the numerator and the denominator tell you in the fractions $\frac{1}{2}$, $\frac{1}{3}$, $\frac{2}{3}$, and $\frac{4}{5}$?

Problem 1.5

The next page shows nine fraction strips of the same length. Each strip is divided into a different number of equal-length parts. On your copy of Labsheet 1.5, label each of the marks on the strips with fraction names in symbolic form. The label for a mark should represent the fraction of the strip to the left of the mark.

Problem 1.5 Follow-Up

Compare these fraction strips with the strips you made in Problem 1.2. Does $\frac{1}{4}$ represent the same length on both strips? Why or why not?

Save your labeled strips so you can use them for the ACE questions and for your work in the next investigation.

Assignment Choices

ACE questions 2–7, 12, and unassigned choices from earlier problems

Answer to Problem 1.5

See page 18k.

Answer to Problem 1.5 Follow-Up

No, $\frac{1}{4}$ does not represent the same length on both strips because the strips are different lengths.

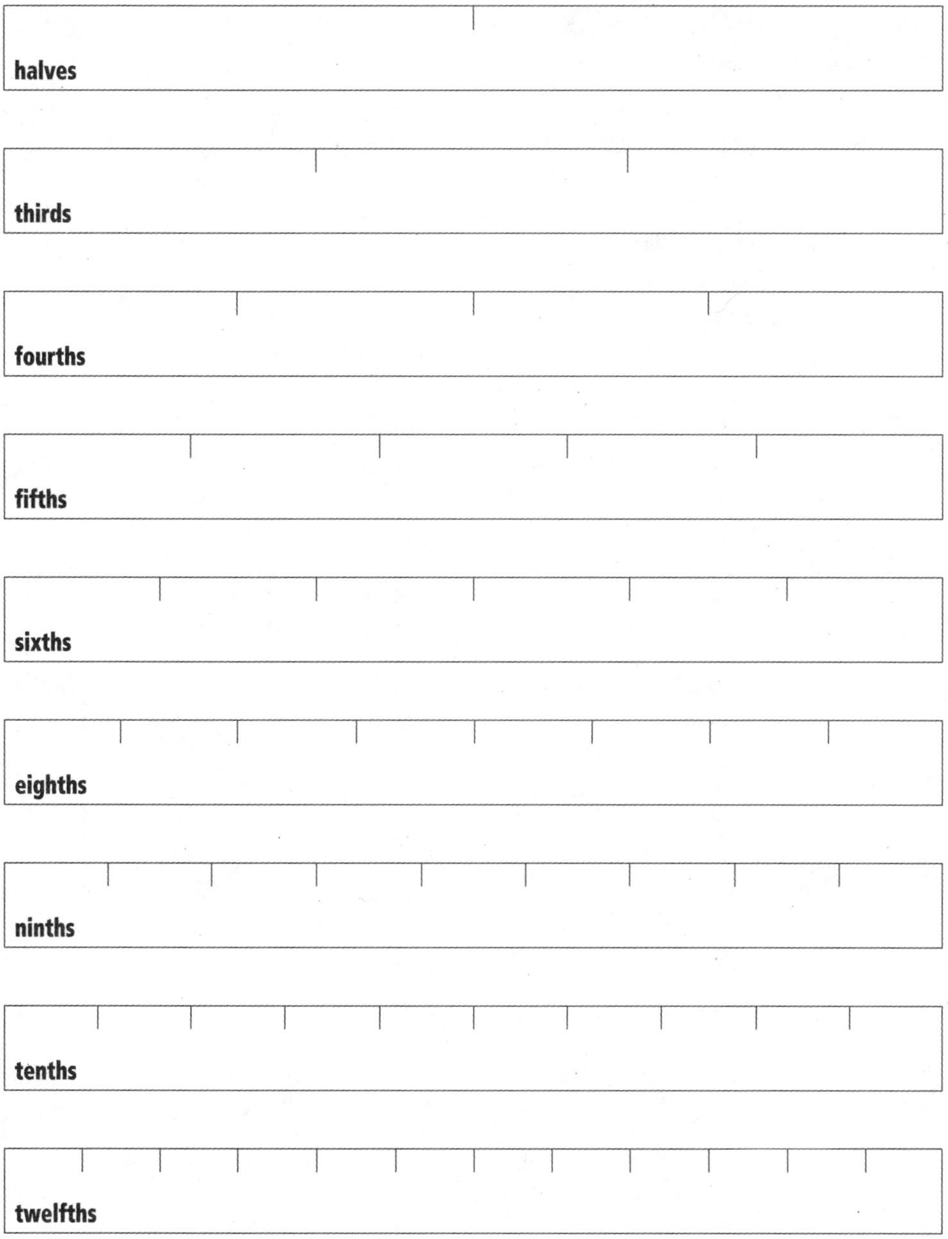

halves

thirds

fourths

fifths

sixths

eighths

ninths

tenths

twelfths

Investigation 1: Fund-Raising Fractions 13

Answers

Applications

1a. The Heron has completed two thirds ($\frac{2}{3}$) of the race and has one third ($\frac{1}{3}$) left to complete. The Palomino has completed one half ($\frac{1}{2}$) of the race and has one half ($\frac{1}{2}$) left to complete. The Flamebird has completed five sixths ($\frac{5}{6}$) of the race and has one sixth ($\frac{1}{6}$) left to complete.

1b. The Heron has completed 400 meters and has 200 meters left to complete. The Palomino has covered 300 meters and has 300 meters left to complete. The Flamebird has covered 500 meters and has 100 meters left to complete.

2.–7. Answers will vary.

As you work on these ACE questions, use your calculator whenever you need it.

Applications

1. At right is a snapshot of three cars drag racing.

    a. For each car, measure from the front of the car to estimate the fraction of the race course completed and the fraction of the race course yet to be covered. You may want to use your fraction strips from Labsheet 1.5.

    b. The drag race course is 600 meters long. For each car, estimate the distance already covered and the distance yet to be covered.

In 2–5, use your fraction strips from Labsheet 1.5 to find something in your home with the given length. Record the name of each object.

2. half ($\frac{1}{2}$) the length of a fraction strip

3. two thirds ($\frac{2}{3}$) the length of a fraction strip

4. one and one half ($1\frac{1}{2}$) times the length of a fraction strip

5. twice the length of a fraction strip

6. Use your fraction strips from Labsheet 1.5 to measure three things in your home that are shorter than a fraction strip. Record the name of each object and its length in terms of a fraction strip.

7. Use your fraction strips from Labsheet 1.5 to measure three things in your home that are longer than a fraction strip. Record the name of each object and its length in terms of a fraction strip.

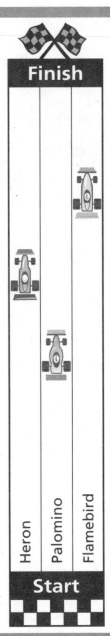

In 8–11, use this drawing of a restaurant drink container. The gauge on the side of the container shows how much of the liquid remains in the container.

8. A full container holds 120 cups.

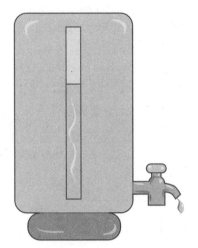

   a. About what fraction of the container is filled with liquid?

   b. About how many cups of liquid are in the container?

   c. About what fraction of the container is empty?

   d. About how many cups of liquid would it take to fill the container?

9. Tell whether each amount is closer to a full container, a half-full container, or an empty container.

   a. five sixths ($\frac{5}{6}$) of a full container

   b. three twelfths ($\frac{3}{12}$) of a full container

   c. five eighths ($\frac{5}{8}$) of a full container

10. About what fraction of the drink container is each of the following amounts?

   a. 37 cups

   b. 10 cups

   c. 55 cups

11. How many pitchers of liquid would it take to fill an empty drink container if a pitcher holds the given amount?

   a. one fourth ($\frac{1}{4}$) of a full container

   b. one third ($\frac{1}{3}$) of a full container

   c. two thirds ($\frac{2}{3}$) of a full container

8a. about two thirds ($\frac{2}{3}$)

8b. about 80 cups

8c. about one third ($\frac{1}{3}$)

8d. about 40 cups

9a. a full container

9b. exactly halfway between empty and half full

9c. one half of a full container

10a. about one third ($\frac{1}{3}$)

10b. one twelfth ($\frac{1}{12}$)

10c. about half ($\frac{1}{2}$)

11a. 4 pitchers

11b. 3 pitchers

11c. one and one half ($1\frac{1}{2}$) pitchers

Investigation 1: Fund-Raising Fractions 15

12a. 4 beetles

12b. 12 beetles

12c. three and one fourth ($3\frac{1}{4}$) fraction strips

Connections

13. six twelfths ($\frac{6}{12}$) or one half ($\frac{1}{2}$)

14. four twelfths ($\frac{4}{12}$) or one third ($\frac{1}{3}$)

15. three twelfths ($\frac{3}{12}$) or one fourth ($\frac{1}{4}$)

16. two twelfths ($\frac{2}{12}$) or one sixth ($\frac{1}{6}$)

17. one twelfth ($\frac{1}{12}$)

18. eight twelfths ($\frac{8}{12}$) or two thirds ($\frac{2}{3}$)

19. nine twelfths ($\frac{9}{12}$) or three fourths ($\frac{3}{4}$)

20. fifteen twelfths ($\frac{15}{12}$), five fourths ($\frac{5}{4}$), or one and one fourth ($1\frac{1}{4}$)

21. eighteen twelfths ($\frac{18}{12}$), three halves ($\frac{3}{2}$), or one and one half ($1\frac{1}{2}$)

22.–25. Answers will vary.

---

12. Ricky found a beetle that was one fourth ($\frac{1}{4}$) the length of a fraction strip from Labsheet 1.5.

  a. How many beetles, placed end to end, would have a total length equal to the length of a fraction strip?

  b. How many beetles, placed end to end, would have a total length equal to three times the length of a fraction strip?

  c. Ricky lined up 13 paper beetles, end to end, each the same length as the one he found. How many times the length of a fraction strip is the length of Ricky's line of beetles?

Connections

In 13–21, tell what fraction of a foot (12 inches) the given length is.

13. 6 inches   **14.** 4 inches   **15.** 3 inches

16. 2 inches   **17.** 1 inch   **18.** 8 inches

19. 9 inches   **20.** 15 inches   **21.** 18 inches

In 22–25, estimate each length as a fraction of a foot.

22. the width of your hand

23. the length of your hand from wrist to fingertip

24. the distance from your wrist to your elbow

25. the distance from your fingertip to your elbow

Extensions

26. Look back at the thermometers on page 7, which show the sixth graders' progress toward their goal.

   a. Make a coordinate graph that shows the sixth-grade fund-raising progress.

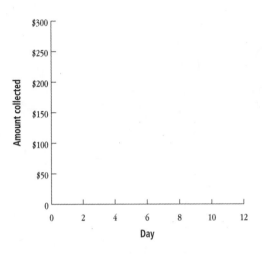

   b. Predict from your graph when the sixth graders would reach their goal if the fund-raiser continued.

   c. Describe the strategy you used to make your prediction.

Extensions

26a. See below left.

26b. Possible answers: day 12, day 13

26c. Possible answer: After 10 days, $250 has been raised, so the students have averaged approximately $25 per day. At this rate, the remaining $50 would be raised in the next two days. However, the graph shows that the amount of money raised per day is declining, so it may take 13 or more days.

26a.

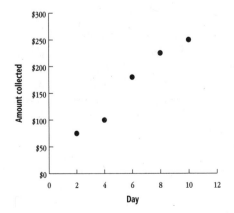

Possible Answers

1. The denominator tells you into how many equal parts the whole has been divided. For example, in the fraction $\frac{2}{3}$, the 3 tells you that the whole has been divided into three equal parts. The numerator tells you how many of those equal parts are being referred to. The 2 in $\frac{2}{3}$ refers to two of the three equal parts.

2. The whole in Problem 1.4 is shorter than the whole in earlier problems, so the fraction strips from Problems 1.2 and 1.3 could not be used *in the same way* they were used for Problems 1.2 and 1.3. Some students may have made new fraction strips; some may have found a different way to use their fraction strips from Problems 1.2 and 1.3; and others may have done Problem 1.4 without fraction strips.

3. The two classes collected the same amount of money only if their dollar goals were the same. If their goals differed, the two classes did not collect the same amount of money. For example, if Ramón's class set a goal of $200 and Melissa's class set a goal of $300, then Ramón's class raised $\frac{3}{5}$ of $200, or $120, and Melissa's class raised $\frac{3}{5}$ of $300, or $180.

Mathematical Reflections

In this investigation, you made fraction strips to help you identify fractional parts of a whole. These questions will help you summarize what you have learned:

① What do the numerator and the denominator of a fraction tell you?

② When you worked on Problem 1.4, did you make new fraction strips? Explain why or why not.

③ Both Ramón's class and Melissa's class reached $\frac{3}{5}$ of their fund-raising goals. Did the two classes raise the same amount of money? Explain your answer.

Think about your answers to these questions, discuss your ideas with other students and your teacher, and then write a summary of your findings in your journal.

Tips for the Linguistically Diverse Classroom

Original Rebus The Original Rebus technique is described in detail in *Getting to Know CMP*. Students make a copy of the text before it is discussed. During discussion, they generate their own rebuses for words they do not understand as the words are made comprehensible through pictures, objects, or demonstrations. Example: Question 1—key words for which students may make rebuses are *numerator* ($\frac{1}{2}$ with the 1 circled), *denominator* ($\frac{1}{2}$ with the 2 circled), *fraction* ($\frac{1}{2}$).

TEACHING THE INVESTIGATION

1.1 • Reporting Our Progress

Launch

In this problem, your students will write announcements describing the progress that a sixth-grade class has made toward their fund-raising goal. This activity lets you quickly assess where your students are in their understanding of fractions as parts of wholes. This problem raises issues about fractions that will be explored in a more systematic way in later problems in the investigation. You should not expect your students to be able to respond in precise mathematical ways to the challenge in this problem.

Your students need to understand the context of this problem. You could begin by asking whether anyone has ever seen a thermometer display showing the progress of a fund-raising effort. Have a student draw an example of such a thermometer on the board, or draw one yourself. Ask questions about the example, focusing attention on the whole. Emphasize that the whole is the distance from the bottom of the thermometer to the goal; it is not necessarily the entire length of the thermometer.

When the class understands the thermometer model, tell them the story of the fund-raising campaigns at Thurgood Marshall School. Have students look at the thermometer on page 5 of the student edition (or display Transparency 1.1). Explain that this thermometer shows the progress of the sixth graders after the second day of their sale.

> Work with your group to write an interesting and informative message describing how the sixth-grade campaign is going. Make sure your announcement tells what part of the $300 goal has been reached and what part remains to be raised.

Explore

Have students work on this problem in pairs or small groups. As groups work, ask questions to help them clarify their messages. Ask them how they drew their conclusions, and encourage them to find other mathematical ways to describe the situation. Be on the lookout for clever ideas and for problems that should be discussed in the summary. You want to keep students focused and to get an idea of who understands the situation and who is confused. You may want to have groups write their results on large sheets of paper or blank transparencies so that they can be shared.

Here are some questions students may ask as they work:

> Do we use fractions or money?

Encourage students to explore both so they understand how the two relate.

> What do you mean by "clever and informative"?

Ask students to think about what they would find interesting to know in such a situation and what they want their announcement to accomplish.

Can I just say a "little bit"?

Explain that one of the nice things about mathematics is that it helps us to convey precise information. A "little bit" is hard to interpret. Encourage the group to be more precise.

How long does the message have to be?

Emphasize that the amount of information the message provides is more important than its length.

Summarize

From your observations of the students, you will have ideas about questions that must arise during the summary and about which groups should be called upon to ensure that important ideas are discussed.

> I am going to choose a few groups to share their messages. After you present your message, explain how you came up with the numbers you used. Those of you listening, note similarities and differences in the strategies and information that the groups report.

Have some or all of the groups read their announcements. After each presentation ask the class some questions.

> Does this announcement make sense?

> Do you agree or disagree? Why or why not?

> What strategy did your group use to reason about the problem?

In one class, a group said that five twelfths of the goal had been reached. The rest of the class agreed that this was an acceptable answer, even though they had all concluded that one fourth, or $75, of the goal had been reached. The teacher did not take the time to correct the students, but instead asked them to continue to think about this as they worked on the next problem.

1.2 • Using Fraction Strips

Launch

In this problem, students make fraction strips by folding strips of paper. Then they use the strips to estimate the fund-raising progress displayed on several thermometers. In Problem 1.5, students will be provided with a set of preprinted fraction strips to ensure accuracy. However, it is essential that they make their own strips first to explore the concept of fractions as parts of wholes.

Provide each student with a set of nine $8\frac{1}{2}$-inch paper strips. Students will make strips to show halves, thirds, fourths, fifths, sixths, eighths, ninths, tenths, and twelfths. The length of each strip is the distance to the goal on the thermometer. If you make the strips 1-inch wide, you can make 11 strips from an $8\frac{1}{2}$-by-11-inch sheet of paper. You will want to have extra strips available for students who make mistakes.

Refer students to page 7 in the student edition, or display Transparency 1.2A. Explain that the thermometers indicate the progress of the sixth-grade fund-raiser after two, four, six, eight, and ten days.

> We can devise a measuring device to help us estimate the progress of the fund-raiser for each of these days.
>
> I have cut some strips that are the same length as the distance from the bottom of the thermometer to the goal. (*Hold up a strip.*) I can make a strip to show halves by folding it into two equal pieces. (*Demonstrate this.*) This fold mark shows me how high the "mercury" must be to indicate one half. I'm going to draw over this fold mark with my pencil so that it is easy to see. (*Demonstrate.*) I can now use this strip to see if any of the thermometers are halfway to the goal.
>
> (*Hold up a new, unfolded strip.*) How can I fold this strip to show thirds?

Fold a strip by following a student's instructions. Ask the class whether they agree that the method results in a strip that is divided into thirds.

> Now, I want you to make your own fraction strips. When everyone is finished, we will discuss what we have found before we move on to the rest of the problem.

Explore

Each student should make a set of fraction strips. Encourage students to label the strips "halves," "thirds," and so on, so that the meaning of each label can be discussed before moving to symbols. It is not necessary for students to label each mark with a fraction, although some of your students may be comfortable enough with symbols to do this.

If students are having difficulty, suggest that they ask another student for help. You may want to establish a rule that students can help each other by giving oral directions, but no student may fold another student's strips.

As students work, circulate and ask questions about what they are doing.

> How can you use the strategy for making the halves strip to help you make the fourths strip?
>
> Did you compare your strips to someone else's to check for accuracy?
>
> Can you explain your folding strategy to another student or to me, so you can see if it makes sense?

Here are some strategies students have used to make their strips.

Halves   Fold a strip in half.

Thirds   Fold a strip into three parts so that the sections are the same length. Or, make an S with a strip and "squish" it together, keeping all three pieces the same size.

Fourths   Fold a strip in half and then in half again.

Fifths   Roll a strip around two fingers two and a half times. Take the strip off your fingers still rolled, and carefully flatten the roll, making the five sections as close as possible to the same length. Or, fold the ends of the strip in toward the middle so that the two sections with overlap and the section in the middle are about the same length. Then make a fold on each side where the overlapping part ends.

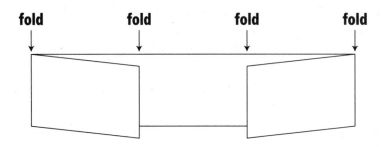

Sixths   Make a thirds strip, then fold each section in half.

Eighths   Make a fourths strip, then fold each section in half.

Ninths   Make a thirds strip, then divide each section into three equal parts.

Tenths   Make a fifths strip, then fold each section in half.

Twelfths Make a sixths strip, then fold each section in half. Or, make a fourths strip, then fold each section into three equal parts.

When most students are finished making their strips, begin a class discussion in which students share their folding strategies. It is important that students spend time connecting fraction words to the fraction-strip model before moving to symbols. Focus on characteristics of the fraction strips that indicate important ideas about fractions. For example, emphasize that the parts into which a strip is folded must be the same size. Point out that you can make strips for both fourths and eighths by starting with a halves strips, and that two fourths and four eighths are the same as one half.

A student may suggest estimating the size of the parts and then tearing off the end of the strip so that the folded pieces are the same length. Here you can emphasize that a set of strips used to compare fractions must represent the *same* whole. In this case, each strip must be exactly $8\frac{1}{2}$ inches long, the distance from the bottom of the thermometer to the goal.

After this discussion, have students work on the rest of the problem, using their strips to describe the progress the sixth graders have made on days 2, 4, 6, 8, and 10 of their fund-raiser.

Summarize

Allow students to share their estimates and strategies. You may want to have them use Transparency 1.2A and a set of fraction strips made from transparency film to explain their answers. It is possible that students will get different estimates; it is hard to be precise with folded fraction strips. Accept estimates that are reasonable.

Focus on word descriptions of fractions. Some students will be able to use symbols from the start. Do not discourage this, but ask them to explain in words so you can assess their understanding.

For the Teacher

Throughout the early problems in this investigation, we encourage you to focus on two ways of representing fractions: visual models (for example, fraction strips) and word names. Since your students may be at a variety of developmental levels in their understanding of fractions, we encourage you to move back and forth between these two representations repeatedly. Research has shown that understanding of fractions is stronger if the use of symbols is delayed until the use of word names is firmly in place. In Problem 1.5 symbols are introduced. At this point, you should begin asking questions that focus on all three representations of fractions—visual models, word names, and symbols. Here are examples of the types of questions you can ask your students to help their understanding:

What is the whole for this situation?

Into how many equal-size parts do you need to divide the whole so that you can model the problem?

What does the numerator of the fraction mean? What does the denominator of the fraction mean?

If two classes reach half their goal, have they raised the same amount of money? Why or why not?

If we make the denominator bigger, what does that say about the size of the parts into which the whole has been divided?

If we make denominator smaller, what does that say about the size of the parts into which the whole has been divided?

If we make the numerator bigger, what does that mean?

If we make the numerator smaller, what does that mean?

You may want to raise the idea of equivalent fractions:

Gloria said she thought the sixth-graders had reached $\frac{1}{4}$ of the goal on day two. Is their another way to name the progress on day two? Is there another fraction strip that has a fold that matches the progress?

You don't need to spend a lot of time discussing equivalent fractions, since we will return to this idea in several investigations, but you don't want to miss the opportunity to have your students begin to search for equivalent fractions.

Continually encourage students to explain and to use more precise language so that everyone can understand what they are sharing. As a result of this discussion, you want your students to be able to describe how to use fraction strips to estimate parts of wholes.

1.3 • Comparing Classes

Launch

In this problem, students compare fractions for which the wholes vary. The sixth-, seventh-, and eighth-grade classes at Thurgood Marshall School have each set a different number of dollars as their fund-raising goal. The problem involves estimating both the fraction of the goal that has been reached and the amount of money that has been raised. This situation is effective because the students easily understand that half of $300 is not as much money as half of $360.

Although the goal amounts for the three classes differ, the lengths of the thermometers are the same. Students use their fraction strips from Problem 1.2 to reason about the part of each whole being shown and then use the goal amounts to determine how much money has been raised. This problem gives students a good sense of the importance of knowing what the whole is when making comparisons.

Refer students to page 9 in the student edition, or show Transparency 1.3A.

> Remember that three classes held fund-raisers. These thermometers show the final results of the fund-raisers for the sixth, seventh, and eighth grades.
>
> At the end of ten days, the seventh graders claimed their class was the most successful. But so did the eighth graders!
>
> Take a minute to examine these thermometers. How are the thermometers alike? How are they different?

You want students to observe that the distance to the goal is the same on all the thermometers. This means that the same fraction strips can be used to measure the progress on each thermometer. However, the dollar amount of the goal is different for each thermometer. This means that the same fraction describes different amounts on each thermometer. So, for example, half of the seventh-grade goal is $200, while half of the eighth-grade goal is only $120. This problem focuses students on the concepts of models and wholes.

> Each grade has a different goal. Look at the sixth-grade thermometer for day 2. We agreed that the class had reached one fourth of its goal by the end of day 2. How much money had been collected?

This conversation may have come up during the summary of Problem 1.1, but you will want to raise it again. Be sure students give a complete explanation for how they determined the amount of money represented. You may have to model a way to think about this:

> If we separate the goal amount of money into four equal parts, how much is each part? Would this amount represent one fourth of the goal? Why or why not?

> In your groups, work on Problem 1.3 and the follow-up. You will need to decide what argument the seventh graders could use to support their claim. Then decide what argument the eighth graders could use. These will probably be different arguments.

Explore

Look for students that are using effective strategies and for students that are having difficulty. Make sure both the strategies and the difficulties are discussed in the summary.

Summarize

Discuss the answers to parts A and B to see whether students have found reasonable answers. Then move to parts C and D. Have each group present its arguments for the seventh and eighth grades. Discuss what is alike and what is different about the arguments presented. This is a good time to focus on the attributes of a convincing argument.

You may want to have students present arguments for all three grade levels. You could also ask them to predict how much longer each fund-raiser would need to continue for each class to reach its goal.

1.4 • Exceeding the Goal

Launch

In Problems 1.2 and 1.3, students were asked to think of the whole both as an amount of money and as the height of the goal mark on a thermometer. Using thermometers that were the same length but which represented different money goals focused students' attention on the importance of considering the whole when interpreting fractions.

Problem 1.4 extends students' thinking in two ways. First, the distance to the goal for the teachers' thermometer is shorter than the distance to the goal for the students' thermometers, so the same fraction strips cannot be used to reason about the teachers' progress. Students will have to create new strips, invent ways to use parts of the strips they already have, or invent a strategy for reasoning without fraction strips. Second, the teachers have exceeded their fund-raising goal, so students must consider fractions that represent more than a whole—that is, fractions greater than 1.

Describe the situation to your students. Refer them to page 11 in the student edition or display Transparency 1.4A. Students should notice that the teachers' thermometer is shorter and that the mercury eventually goes beyond the goal mark, indicating that the teachers have exceeded their goal. You may want them to compare the students' thermometers with the teachers' thermometer to observe what is alike and different about them.

Tell students that during the summary you will expect them to tell about the strategies they used to solve the problem and to explain why they think their announcement is an accurate portrayal of the teachers' progress.

Explore

Have students work in groups on Problem 1.4 and the follow-up. Circulate as they work, noting creative or interesting strategies that should be shared with the class.

Although the $8\frac{1}{2}$-inch strips are difficult to use for this problem, some students may make use of them in some way. Question these students about how they are reasoning. In one class, a student said the sixths strip could be used because four sixths of it equaled the distance to the teachers' goal. He explained that the thermometer for day 2 represented one fourth of the teachers' goal, because the mercury was the length of one of the four sixths sections needed to make the whole thermometer. He reasoned that the thermometer for day 6 represented 100% of the goal because the mercury went up to the goal mark, and that the thermometer for day 10 represented one and one fourth of the goal because it was five sixths of his strip—four sixths represented a whole, and the other sixth represented one fourth of the way to another goal.

Some students will want to make new strips. Have some strips cut to the length of the distance to the teachers' goal in case a group decides they need to fold new strips to represent the problem. The students do not need to make new strips if they find other strategies to solve the problem.

Summarize

Have students share strategies for working with the new whole and present the announcements they wrote for the follow-up. Make sure they explain why they think their strategy is reasonable. This will tell you much about their understanding of the part-whole meaning of fractions. You may want to give groups a few minutes to revise their announcements after the class discussion of strategies.

Following are additional questions to stretch the class's thinking:

> The sixth graders' goal was $300. Where would $300 be on the teachers' thermometer?

Students will need to determine what part $300 is of $360. Some might see that if $360 is thought of as six groups of $60, then $300 is five of these groups. So $300 would be represented by a mark five sixths of the way to the goal mark. You might remind students that they know how to analyze whole numbers to see what they have in common. Knowing the greatest common factor is helpful in reasoning about this problem.

> Why is $300 at a different height on the teachers' thermometer than on the sixth graders' thermometer?

How high would one half be on the seventh graders' thermometer? On the teachers' thermometer? Why are the lengths different?

If fractions tell parts of a whole, how do we talk about more than a whole?

Fractions tell us how to talk about any number of wholes and parts of wholes. The teachers' thermometer for day 10 shows one complete whole plus one fourth of another whole. (That is what $1\frac{1}{4}$, or $\frac{5}{4}$, means.)

1.5 • Using Symbolic Form

Launch

Up to this point, your students have worked with fraction strips they folded themselves. The intent has been to develop a secure sense of meaning for parts of the strips. In this problem, students are asked to make the connection between parts of fraction strips and the symbols that represent them, helping them to understand what the numerator and denominator stand for in the symbolic representation of fractions.

After they complete this problem, students will know three ways to represent fractions: concrete models, word names, and symbols. It is important for you to cycle through these three representations repeatedly so that your students make strong connections among them and develop the ability to move flexibly between representations.

Discuss with your students the precision of their folded strips.

> The fraction strips you folded gave you a pretty good way to measure the fund-raising progress, but most are not accurate enough for work that requires more precision—such as accurately comparing fractions or finding fractions that represent the same quantity.

> I will give each of you a labsheet with printed fraction strips. These strips were created with a computer and will allow you to measure more accurately.

Pass out a copy of Labsheet 1.5 to each student. Note that these strips are 16 cm long, not $8\frac{1}{2}$ inches long like the hand-folded strips.

Talk about what the numerator and denominator of a fraction are, focusing on the part-whole interpretation. The *denominator* tells us into how many *equal-size* parts the whole has been divided, and the *numerator* tells us how many of these *equal-size* parts represent the quantity to which we are referring.

Explore

Circulate as students work in pairs to label their page of fraction strips, asking questions about what the symbols mean. Have students refer to their strips to show you what some of the fractions they wrote mean. As you look at a student's fractions strips, you can ask which of two fractions on a strip represents the greater amount. You can also begin to pose questions about different symbol names for a particular length. These ideas of *comparing* fractions and finding *equivalent* fractions will be central themes in many of the remaining investigations.

Summarize

Ask students to compare the fraction strips they folded to the preprinted strips.

> Is one half the same in both sets of strips? Why or why not?

Students should understand that one half does not represent the same length on both strips because the strips are different lengths. However, in each case the symbol notation $\frac{1}{2}$ represents one of two equal-size parts of a whole.

You may want to show a transparent fraction strip and a transparent centimeter ruler on the overhead projector.

> How do our new fraction strips compare to a centimeter ruler? How are they alike? How are they different?

The centimeter ruler has marks that align with marks on many of the fraction strips. For example, on the fourths strip, the first mark is at 4 cm, the second is at 8 cm, the third is at 12 cm, and the whole is at 16 cm. However, the marks on the ninths strip do not align with centimeter marks on the ruler.

Students should save these fractions strips. These strips will be used in the ACE questions and in future investigations. You may want to have students destroy their folded strips to prevent confusion about which fraction strips the investigations that follow involve.

Additional Answers

Answer to Problem 1.1 Follow-Up

Answers will vary. Students may have used a ruler to measure the length of the "mercury" and then compared this length to the total distance to the goal (the distance to the goal is $8\frac{1}{2}$ inches, and the thermometer registers $2\frac{1}{8}$ inches). The distance to the goal is the width of a sheet of paper, so students may have folded a sheet of notebook paper to determine their answers.

Answer to Problem 1.5

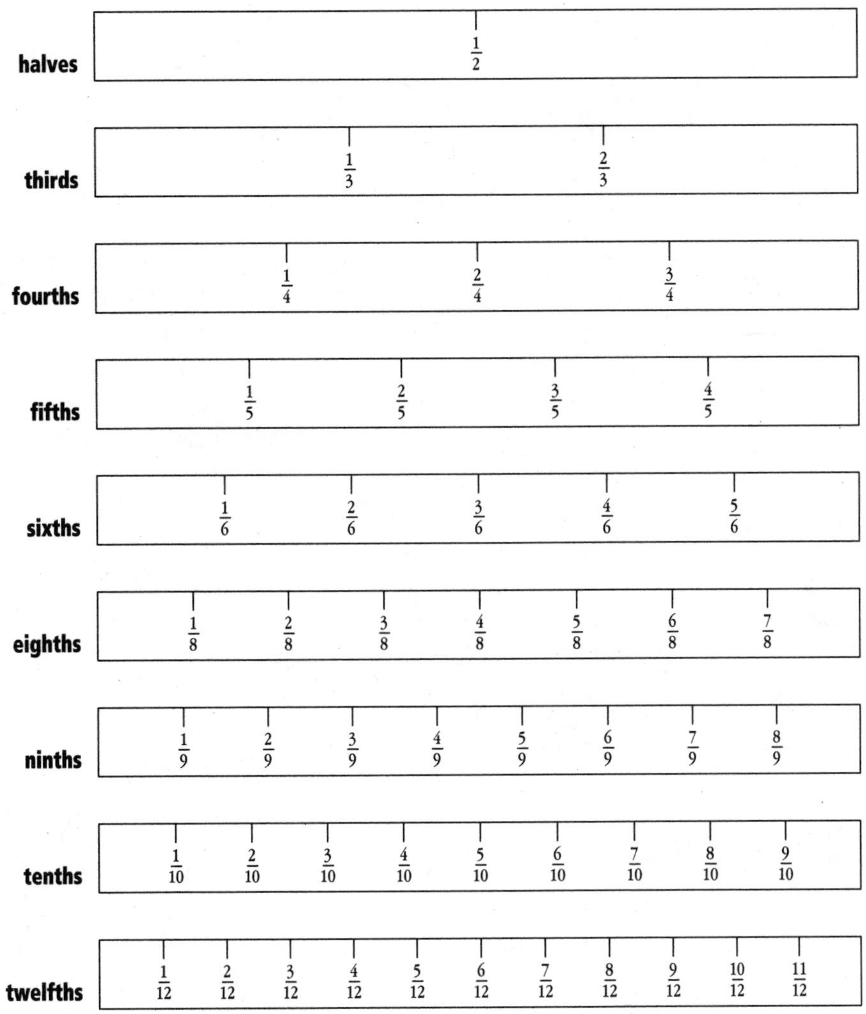

halves

$\frac{1}{2}$

thirds

$\frac{1}{3}$ $\frac{2}{3}$

fourths

$\frac{1}{4}$ $\frac{2}{4}$ $\frac{3}{4}$

fifths

$\frac{1}{5}$ $\frac{2}{5}$ $\frac{3}{5}$ $\frac{4}{5}$

sixths

$\frac{1}{6}$ $\frac{2}{6}$ $\frac{3}{6}$ $\frac{4}{6}$ $\frac{5}{6}$

eighths

$\frac{1}{8}$ $\frac{2}{8}$ $\frac{3}{8}$ $\frac{4}{8}$ $\frac{5}{8}$ $\frac{6}{8}$ $\frac{7}{8}$

ninths

$\frac{1}{9}$ $\frac{2}{9}$ $\frac{3}{9}$ $\frac{4}{9}$ $\frac{5}{9}$ $\frac{6}{9}$ $\frac{7}{9}$ $\frac{8}{9}$

tenths

$\frac{1}{10}$ $\frac{2}{10}$ $\frac{3}{10}$ $\frac{4}{10}$ $\frac{5}{10}$ $\frac{6}{10}$ $\frac{7}{10}$ $\frac{8}{10}$ $\frac{9}{10}$

twelfths

$\frac{1}{12}$ $\frac{2}{12}$ $\frac{3}{12}$ $\frac{4}{12}$ $\frac{5}{12}$ $\frac{6}{12}$ $\frac{7}{12}$ $\frac{8}{12}$ $\frac{9}{12}$ $\frac{10}{12}$ $\frac{11}{12}$

Comparing Fractions

Mathematical and Problem-Solving Goals

- **To continue to use fraction strips as tools for understanding fraction concepts**

- **To investigate the concepts of comparison and equivalence of fractions**

- **To use fractions that are less than, equal to, and greater than 1**

- **To apply knowledge gained by using fraction strips to name, estimate, and compare fractions and to find equivalent fractions**

- **To build a number line and label points between whole numbers**

Student Pages	19–30
Teaching the Investigation	30a–30k

In this investigation, students continue to use fraction strips as tools for understanding fractions. The problems in this investigation were designed to develop an understanding of equivalence and comparison of fractions.

Problem 2.1, Comparing Notes, asks students to investigate competing claims of three teachers about their fund-raising progress. Two of the teachers' claims turn out to be the same, raising the issue of equivalent fractions. In Problem 2.2, Finding Equivalent Fractions, students are asked to find other names for $\frac{2}{3}$ and $\frac{3}{4}$ by comparing fraction strips. They use the patterns they discover to find equivalent fractions for $\frac{1}{8}$, $\frac{2}{5}$, and $\frac{5}{6}$. In Problem 2.3, Making a Number Line, students transfer the fractions from all of their fraction strips onto a single number line. This helps them make sense of the number line and the numbers—fractions—used to label the points between whole numbers. In Problem 2.4, Comparing Fractions to Benchmarks, students use the benchmark values of 0, $\frac{1}{2}$, and 1 to estimate the size of fractions and to compare fractions. Problem 2.5, Fractions Greater Than One, asks students to consider fractions greater than 1 on the number line. As students label points between 1 and 2, they should begin to think about how many fractions are between 1 and 2. Students need time to become acquainted with this notion of density: between any two fractions, there is another fraction.

Materials

For students

- Labeled fraction strips from Labsheet 1.5

For the teacher

- Transparencies 2.1, 2.2A, 2.2B, 2.3, 2.4, and 2.5 (optional)

- A large number line to display in the classroom (see Problems 2.4 and 2.5)

- Fraction strips for the overhead projector (optional; copy Labsheet 1.5 onto blank transparency film)

- Index cards (optional)

INVESTIGATION 2

Comparing Fractions

In Investigation 1, you made fraction strips to help you determine what fraction of the fund-raising goal students had reached. You learned to interpret fractions as parts of a whole. In this investigation, you will look at situations in which you need to compare fractions. Your fraction strips will be a useful model to help you make comparisons.

2.1 Comparing Notes

At the end of the fourth day of their fund-raising campaign, the teachers at Thurgood Marshall School had raised $270 of the $360 they needed to reach their goal. Three of the teachers got into a debate about how they would report their progress.

- Ms. Mendoza wanted to announce that the teachers had made it three fourths of the way to their goal.

- Mr. Park said that six eighths was a better description.

- Ms. Christos suggested that two thirds was really the simplest way to describe the teachers' progress.

Problem 2.1

A. Which of the three teachers do you agree with? Why?

B. How could the teacher you agreed with in part A prove his or her case?

■ **Problem 2.1 Follow-Up**

Name another fraction that describes the teachers' progress.

Goal — $360

Day 4

Teachers'
Book Sale

Answers to Problem 2.1

A. Ms. Mendoza and Mr. Park are correct, since three fourths of $360 and six eighths of $360 are both equal to $270. Ms. Christos is incorrect, since two thirds of $360 is $240.

B. Ms. Mendoza could say that if you divide $360 into four equal parts, each part would be $90 and three of these parts would be $270; three fourths of $360 is $270. Mr. Park could say that if you divide $360 into eight equal parts, each part would be $45 and six of these parts would be $270; six eighths of $360 is $270.

Answer to Problem 2.1 Follow-Up

Possible answer: The teachers could say they had reached $\frac{9}{12}$ of their goal.

At a Glance

Launch

- Review with students the story of the teachers' debate.

Explore

- Using a think-pair-share strategy, have students explore the problem.

- Watch for students with particularly creative reasoning or with ideas that need further development.

Summarize

- Have several pairs share their results.

Assignment Choices

Unassigned choices from earlier problems

2.2

Finding Equivalent Fractions

Launch

- Review what students discovered about $\frac{3}{4}$ and $\frac{6}{8}$ in Problem 2.1.

- Help students understand Problem 2.2.

Explore

- Visit groups as they work, asking questions to guide them in exploring equivalent fractions.

Summarize

- Have several groups share the patterns they discovered for finding equivalent fractions.

- Ask questions to help students discover the pattern: multiply the numerator and denominator by the same number.

- Use visual tools to help students understand that when they multiply the numerator and denominator by the same number they are cutting each of the pieces into which the whole has been divided into smaller pieces.

2.2 Finding Equivalent Fractions

As you worked with your fraction strips, you found that some quantities can be described by several different fractions. In fact, *any* quantity can be described by an infinite number of different fractions!

> ### Did you know?
>
> Hieroglyphic inscriptions from more than 4000 years ago indicate that, with the exception of $\frac{2}{3}$, Egyptian mathematicians used only fractions with 1 in the numerator. Such fractions are known as *unit fractions.* Other fractions were expressed as sums of these unit fractions. The fraction $\frac{2}{7}$, for example, was expressed as $\frac{1}{4} + \frac{1}{28}$.

Two fractions that name the same quantity are called **equivalent fractions.** For example, you probably know several names for the quantity $\frac{1}{2}$. As long as the whole is the same, $\frac{1}{2}$ means the same as $\frac{2}{4}, \frac{3}{6}, \frac{4}{8}, \frac{5}{10}$, and so on. You can show this with fraction strips.

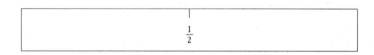

ACE questions 1–7, 19, 38–40, and unassigned choices from earlier problems

Answers to Problem 2.2

A.   The fractions shown on the strips are $\frac{4}{6}$, $\frac{6}{9}$, and $\frac{8}{12}$. Additional equivalent fractions include $\frac{10}{15}$, $\frac{12}{18}$, and $\frac{14}{21}$.

B.   The fractions shown on the strips are $\frac{6}{8}$, $\frac{9}{12}$, and $\frac{12}{16}$. Additional equivalent fractions include $\frac{15}{20}$, $\frac{18}{24}$, and $\frac{21}{28}$.

C.   Answers will vary. Students may realize that you can find equivalent fractions by multiplying the numerator and the denominator by the same number.

Problem 2.2

The fraction strips on the left below show $\frac{2}{3}$ and three fractions equivalent to $\frac{2}{3}$. The strips on the right show $\frac{3}{4}$ and three fractions equivalent to $\frac{3}{4}$. Study the two sets of strips. Look for patterns that will help you find other equivalent fractions.

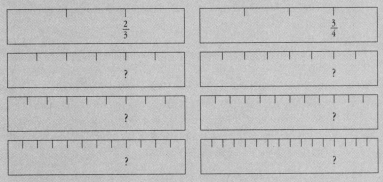

A. What are the three fractions shown that are equivalent to $\frac{2}{3}$? Name three more fractions that are equivalent to $\frac{2}{3}$.

B. What are the three fractions shown that are equivalent to $\frac{3}{4}$? Name three more fractions that are equivalent to $\frac{3}{4}$.

C. What pattern do you see that can help you find equivalent fractions?

■ Problem 2.2 Follow-Up

Test your ideas by naming at least five fractions equivalent to each given fraction.

1. $\frac{1}{8}$

2. $\frac{2}{5}$

3. $\frac{5}{6}$

Answers to Problem 2.2 Follow-Up

1. Possible answers: $\frac{2}{16}, \frac{3}{24}, \frac{4}{32}, \frac{5}{40}, \frac{6}{48}, \frac{7}{56}$

2. Possible answers: $\frac{4}{10}, \frac{6}{15}, \frac{8}{20}, \frac{10}{25}, \frac{12}{30}, \frac{14}{35}$

3. Possible answers: $\frac{10}{12}, \frac{15}{18}, \frac{20}{24}, \frac{25}{30}, \frac{30}{36}, \frac{35}{42}$

2.3

Making a Number Line

At a Glance

Launch

■ Demonstrate how to transfer marks from fraction strips to a number line.

Explore

■ Have students create their own number lines.

■ Have students work on the follow-up to add numbers that are not represented on their fraction strips to their number lines.

Summarize

■ Display several students' number lines.

■ Focus students on using their number lines to compare two fractions by finding equivalent fractions with the same denominator.

It would be helpful to have one strip that shows all of the fractions in your set of fraction strips. That way, you could measure fractional lengths using only one strip.

To make this master strip, you can copy the fractions from all of your fraction strips onto a single number line. The result will be a number line from 0 to 1 with marks for $\frac{1}{4}$, $\frac{1}{3}$, $\frac{1}{2}$, $\frac{2}{4}$, $\frac{2}{3}$, $\frac{3}{4}$, and all the other fractions on your strips.

Here's one way to transfer all of the fractions from your fraction strips onto one number line.

1. Draw a line at the top of a sheet of paper. This will be your number line. Label the left end of the number line with the numeral 0. Line up one end of a fraction strip with this 0 point, and make a mark where the other end crosses the number line. Label this mark with the numeral 1.

2. Align the end of your halves fraction strip (from Labsheet 1.5) with the 0 mark. Make a mark where the $\frac{1}{2}$ mark crosses the number line. Label this mark with the fraction $\frac{1}{2}$.

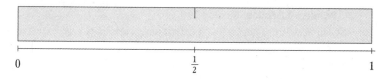

3. Align the end of your thirds fraction strip with the 0 mark. Make and label marks where the $\frac{1}{3}$ and $\frac{2}{3}$ marks cross the number line.

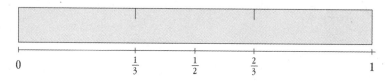

4. Continue this process with the rest of your strips.

Assignment Choices

ACE questions 20–27, 37, and unassigned choices from earlier problems

Answers to Problem 2.3

A. See page 30j.

B. Answers will vary.

Answers to Problem 2.3 Follow-Up

Answers will vary.

Comparing Fractions to Benchmarks

Problem 2.3

A. Make a number line as described above. When you find another name for a mark you have already labeled, record the new name below the first name.

B. Look for patterns in your finished number line. Record your findings.

■ **Problem 2.3 Follow-Up**

Mark and label three fractions on your number line that are not represented on your set of fraction strips.

2.4 Comparing Fractions to Benchmarks

When you solve problems involving fractions, you may find it useful to estimate the size of fractions quickly. One strategy is to compare each fraction to 0, $\frac{1}{2}$, and 1. These values serve as **benchmarks**—or reference points. First, you can decide whether a fraction is between 0 and $\frac{1}{2}$, between $\frac{1}{2}$ and 1, or greater than 1. Then you can decide whether the fraction is closest to 0, $\frac{1}{2}$, or 1.

Problem 2.4

A. Decide whether each fraction below is between 0 and $\frac{1}{2}$ or between $\frac{1}{2}$ and 1.

$\frac{1}{5}$ $\frac{2}{3}$ $\frac{8}{10}$ $\frac{3}{12}$ $\frac{3}{5}$ $\frac{5}{6}$ $\frac{5}{8}$ $\frac{4}{5}$ $\frac{3}{8}$ $\frac{3}{4}$ $\frac{2}{9}$ $\frac{7}{12}$ $\frac{1}{3}$

B. Decide whether each fraction from part A is closest to 0, $\frac{1}{2}$, or 1. Record your information in a table.

C. Explain your strategies for comparing fractions to 0, $\frac{1}{2}$, and 1.

D. Use benchmarks and other strategies to help you write the fractions from part A in order from smallest to largest.

■ **Problem 2.4 Follow-Up**

1. In a–d, decide which fraction is larger by using benchmarks or another strategy that makes sense to you. Then write each pair of fractions, inserting a less-than symbol (<), a greater-than symbol (>), or an equals symbol (=) between the fractions to make a true statement. Describe your reasoning.

a. $\frac{3}{12}$ $\frac{7}{12}$ **b.** $\frac{5}{6}$ $\frac{5}{8}$

c. $\frac{2}{3}$ $\frac{3}{9}$ **d.** $\frac{13}{12}$ $\frac{6}{5}$

At a Glance

Launch

■ Demonstrate how to determine whether a fraction is closest to 0, $\frac{1}{2}$, or 1.

Explore

■ Circulate as groups work, asking students to explain their strategies for determining which benchmark each fraction is near and for ordering the fractions.

Summarize

■ Have students share their strategies for categorizing and ordering the fractions.

■ As a class, further develop the concepts raised in each part of the problem.

■ Discuss the exercises in the follow-up.

Answers to Problem 2.4

A. See the lists in the "Summarize" section on page 30e.

B. See the table in the "Summarize" section on page 30f.

C. Answers will vary.

D. $\frac{1}{5}, \frac{2}{9}, \frac{3}{12}, \frac{1}{3}, \frac{3}{8}, \frac{7}{12}, \frac{3}{5}, \frac{5}{8}, \frac{2}{3}, \frac{3}{4}, \frac{8}{10} = \frac{4}{5}, \frac{5}{6}$

Answers to Problem 2.4 Follow-Up

See page 30j.

Assignment Choices

ACE questions 8–15, 41, 42, and unassigned choices from earlier problems

Fractions Greater Than One

At a Glance

Launch

- With the class, explore the section of the number line between 1 and 2.

- Discuss how to label points with both mixed numbers and improper fractions.

Explore

- Circulate as pairs label the lettered points on the number line, asking questions to help them understand improper fractions.

- Have pairs locate points fitting the constraints in part B.

Summarize

- Have students share their strategies for labeling points.

- If students are ready, discuss an efficient strategy for converting between mixed numbers and improper fractions.

---

2. In a–f, use your fraction strips or another method to compare the fractions in each pair. Then write each pair of fractions, inserting <, >, or = between the fractions to make a true statement. Describe your reasoning.

a. $\frac{4}{7}$ $\frac{6}{7}$ b. $\frac{7}{10}$ $\frac{8}{12}$

c. $\frac{5}{8}$ $\frac{6}{9}$ d. $\frac{10}{12}$ $\frac{5}{6}$

e. $\frac{2}{4}$ $\frac{5}{9}$ f. $\frac{3}{9}$ $\frac{3}{10}$

2.5 Fractions Greater Than One

The whole-number points on a number line follow one another in a simple, regular pattern. But, as you saw in Problem 2.3, between every pair of whole numbers are many other points that may be labeled with fractions.

The portion of the number line shown below has marks for halves, thirds, fourths, fifths, sixths, eighths, ninths, tenths, and twelfths. These marks are different from the marks you identified in Problem 2.3, because they indicate fractions that are between 1 and 2 instead of between of 0 and 1.

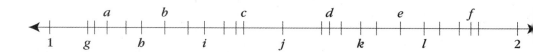

Problem 2.5

A. Use the fraction strips from Labsheet 1.5 to find as many labels as you can for each of the lettered points. For each point, record the letter and the fraction labels.

B. Copy the number line onto a sheet of paper. Mark and label a point fitting each description below. Do not use points that are already marked.

1. a point close to, but larger than, 1

2. a point close to, but smaller than, $1\frac{1}{2}$

3. a point close to, but larger than, $1\frac{1}{2}$

4. a point close to, but smaller than, 2

Assignment Choices

ACE questions 16–18, 28–36, and unassigned choices from earlier problems

Assessment

It is appropriate to use Check-Up 1 after this problem.

Answers to Problem 2.5

A.   a. $1\frac{1}{8}$ and $\frac{9}{8}$ b. $1\frac{1}{4}$, $1\frac{2}{8}$, $1\frac{3}{12}$, $\frac{5}{4}$, $\frac{10}{8}$, and $\frac{15}{12}$

   c. $1\frac{5}{12}$ and $\frac{17}{12}$ d. $1\frac{3}{5}$, $1\frac{6}{10}$, $\frac{8}{5}$, and $\frac{16}{10}$

   e. $1\frac{3}{4}$, $1\frac{6}{8}$, $1\frac{9}{12}$, $\frac{7}{4}$, $\frac{14}{8}$, and $\frac{21}{12}$ f. $1\frac{9}{10}$ and $\frac{19}{10}$

   g. $1\frac{1}{12}$ and $\frac{13}{12}$ h. $1\frac{1}{5}$, $1\frac{2}{10}$, $\frac{6}{5}$, and $\frac{12}{10}$

   i. $1\frac{1}{3}$, $1\frac{2}{6}$, $1\frac{3}{9}$, $1\frac{4}{12}$, $\frac{4}{3}$, $\frac{8}{6}$, $\frac{12}{9}$, and $\frac{16}{12}$ j. $1\frac{1}{2}$, $1\frac{2}{4}$, $1\frac{3}{6}$, $1\frac{4}{8}$, $1\frac{5}{10}$, $1\frac{6}{12}$, $\frac{3}{2}$, $\frac{6}{4}$, $\frac{12}{8}$, $\frac{15}{10}$, and $\frac{18}{12}$

   k. $1\frac{2}{3}$, $1\frac{4}{6}$, $1\frac{6}{9}$, $1\frac{8}{12}$, $\frac{5}{3}$, $\frac{10}{6}$, $\frac{15}{9}$, and $\frac{20}{12}$ l. $1\frac{4}{5}$, $1\frac{8}{10}$, $\frac{9}{5}$, and $\frac{18}{10}$

B.   1. Possible answer: $1\frac{1}{16}$

   2. Possible answer: $1\frac{7}{16}$

   3. Possible answer: $1\frac{9}{16}$

   4. Possible answer: $1\frac{15}{16}$

▓ Problem 2.5 Follow-Up

Find an equivalent fraction, with a denominator greater than 12, for one of the lettered points. Explain how you arrived at your answer.

Answers to Problem 2.5 Follow-Up

Answers will vary.

Answers

Applications

1.–4. See below right.

5. $\frac{3}{4}, \frac{6}{8}, \frac{9}{12}$

6. See page 30j.

7. See page 30j.

8. $\frac{8}{10} > \frac{3}{8}$

9. $\frac{2}{3} > \frac{4}{9}$

10. $\frac{3}{5} > \frac{5}{12}$

11. $\frac{1}{3} < \frac{2}{3}$

As you work on these ACE questions, use your calculator whenever you need it.

Applications

In 1–4, decide whether the statement is true or false. Explain your reasoning in words or by drawing pictures.

1. $\frac{1}{3} = \frac{4}{12}$ **2.** $\frac{4}{6} = \frac{2}{3}$ **3.** $\frac{2}{5} = \frac{1}{3}$ **4.** $\frac{2}{4} = \frac{5}{10}$

5. The drawing below shows the volume indicator on a stereo receiver. Use the fraction strips shown to find three fractions that describe the part of the maximum volume shown by the indicator.

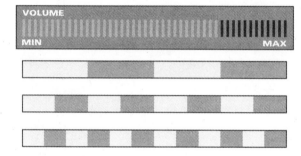

In 6 and 7, draw fraction strips to show that the two fractions are equivalent.

6. $\frac{2}{5}$ and $\frac{6}{15}$ **7.** $\frac{1}{9}$ and $\frac{2}{18}$

In 8–15, use benchmarks or another strategy that makes sense to you to decide which fraction is larger. Then write each pair of fractions, inserting a less-than symbol (<), a greater-than symbol (>), or an equals symbol (=) between the fractions to make a true statement. Describe your reasoning.

8. $\frac{8}{10}$ $\frac{3}{8}$ **9.** $\frac{2}{3}$ $\frac{4}{9}$

10. $\frac{3}{5}$ $\frac{5}{12}$ **11.** $\frac{1}{3}$ $\frac{2}{3}$

1. true; equals

2. true; equals

3. false; does not equal

4. true; equals

12. $\frac{3}{4}$ $\frac{3}{5}$ **13.** $\frac{3}{2}$ $\frac{7}{6}$

14. $\frac{8}{12}$ $\frac{6}{9}$ **15.** $\frac{9}{10}$ $\frac{10}{11}$

16. Describe, in writing or with pictures, how $\frac{7}{3}$ compares with $2\frac{1}{3}$.

17. Which is larger, $\frac{7}{6}$ or $\frac{13}{12}$? Explain your reasoning.

18. On the number line from 0 to 10, where is $\frac{13}{3}$ located? Explain your reasoning.

19. Write an explanation to a friend of how to find a fraction that is equivalent to $\frac{3}{5}$. You can use words and pictures to help explain.

Connections

In 20–25, copy each number line, and then estimate and mark where the numeral 1 would be.

20.

21.

22.

23.

20.

21.

---

12. $\frac{3}{4} > \frac{3}{5}$

13. $\frac{3}{2} > \frac{7}{6}$

14. $\frac{8}{12} = \frac{6}{9}$

15. $\frac{9}{10} < \frac{10}{11}$

16. $\frac{7}{3}$ and $2\frac{1}{3}$ are equivalent. Since 3 thirds is equal to 1 whole, 7 thirds equals 2 wholes plus a third, or $2\frac{1}{3}$.

17. $\frac{7}{6} > \frac{13}{12}$; $\frac{7}{6}$ is equal to $\frac{14}{12}$, and $\frac{14}{12}$ is greater than $\frac{13}{12}$.

18. $\frac{13}{3}$ falls one third of the way from 4 to 5 on the number line. Since 3 thirds is one whole, 13 thirds is 4 wholes plus a third, or $4\frac{1}{3}$.

19. Possible answer: You could draw a fraction strip and divide it into five equal parts. Shade three of these parts to represent $\frac{3}{5}$. Then divide each of the five parts into two equal parts. You would then have ten equal parts, and six of the parts would be shaded. Therefore, $\frac{3}{5}$ is the same as $\frac{6}{10}$, so $\frac{6}{10}$ is equivalent to $\frac{3}{5}$.

Connections

20. See below left.

21. See below left.

22. See page 30k.

23. See page 30k.

24. See below right.

25. See below right.

26. $\frac{1}{3}$

27. $\frac{5}{6}$

24.

25.

In 26 and 27, write a fraction to describe the part of the length of a new pencil represented by the old pencil.

26.

27.

In 28–36, compare each fraction to the benchmarks 0, $\frac{1}{2}$, 1, $1\frac{1}{2}$, and 2. Determine between which two benchmarks the fraction falls, and then determine to which benchmark the fraction is nearest. Organize your answers in a table. The columns of your table should be labeled as shown here:

Number	Lower benchmark	Upper benchmark	Nearest benchmark

24.

25.

28. $\frac{3}{5}$ **29.** $1\frac{2}{6}$ **30.** $\frac{12}{10}$

31. $\frac{2}{18}$ **32.** $1\frac{8}{10}$ **33.** $1\frac{1}{10}$

34. $\frac{12}{24}$ **35.** $\frac{9}{6}$ **36.** $1\frac{12}{15}$

37. These bars represent trips that Ms. Axler took in her job this week.

300 km ☐

180 km ☐

240 km ☐

   a. On a copy of each bar, shade in the distance Ms. Axler had traveled when she had gone one third of the total distance for the trip.

   b. How many kilometers had Ms. Axler traveled when she was at the one-third point in each trip? Explain your reasoning.

Extensions

In 38–40, find every fraction with a denominator less than 50 that is equivalent to the given fraction.

38. $\frac{3}{15}$ **39.** $\frac{8}{3}$ **40.** $1\frac{4}{6}$

41. Find five fractions between $\frac{1}{4}$ and $\frac{1}{2}$.

42. Which of the fractions below represents the largest part of a whole? Explain your reasoning.

$$\frac{4}{5} \qquad \frac{17}{23} \qquad \frac{51}{68}$$

28.–36. See below left.

37a. See page 30k.

37b. 100 km, 60 km, 80 km; Explanations will vary.

Extensions

38. $\frac{6}{30}, \frac{9}{45}$

39. $\frac{16}{6}, \frac{24}{9}, \frac{32}{12}, \frac{40}{15}, \frac{48}{18}, \frac{56}{21},$ $\frac{64}{24}, \frac{72}{27}, \frac{80}{30}, \frac{88}{33}, \frac{96}{36}, \frac{104}{39}, \frac{112}{42},$ $\frac{120}{45}, \frac{128}{48}$

40. $\frac{10}{6}, \frac{20}{12}, \frac{30}{18}, \frac{40}{24}, \frac{50}{30}, \frac{60}{36},$ $\frac{70}{42}, \frac{80}{48}$

41. Possible answer: $\frac{5}{16},$ $\frac{3}{8}, \frac{13}{32}, \frac{7}{16}, \frac{31}{64}$

42. $\frac{4}{5}$; To compare $\frac{4}{5}$ to $\frac{17}{23}$, you can write equivalent fractions with 5×23, or 115, as the denominator. This gives $\frac{4}{5} = \frac{92}{115}$ and $\frac{17}{23} = \frac{85}{115}$. So $\frac{4}{5}$ is larger. To compare $\frac{4}{5}$ and $\frac{51}{68}$, you can use 5×68, or 340, as the common denominator to get $\frac{4}{5} = \frac{272}{340}$ and $\frac{51}{68} = \frac{255}{340}$. So $\frac{4}{5}$ is larger.

	Number	Lower benchmark	Upper benchmark	Nearest benchmark
28.	$\frac{3}{5}$	$\frac{1}{2}$	1	$\frac{1}{2}$
29.	$1\frac{2}{6}$	1	$1\frac{1}{2}$	$1\frac{1}{2}$
30.	$\frac{12}{10}$	1	$1\frac{1}{2}$	1
31.	$\frac{2}{18}$	0	$\frac{1}{2}$	0
32.	$1\frac{8}{10}$	$1\frac{1}{2}$	2	2
33.	$1\frac{1}{10}$	1	$1\frac{1}{2}$	1
34.	$\frac{12}{24}$	$\frac{1}{2}$	$\frac{1}{2}$	equal to $\frac{1}{2}$
35.	$\frac{9}{6}$	$1\frac{1}{2}$	$1\frac{1}{2}$	equal to $1\frac{1}{2}$
36.	$1\frac{12}{15}$	$1\frac{1}{2}$	2	2

Possible Answers

1. $\frac{6}{10}$, $\frac{9}{15}$, $\frac{12}{20}$, $\frac{15}{25}$, $\frac{18}{30}$, and $\frac{21}{35}$; These fractions can be found by multiplying the numerator and denominator of $\frac{3}{5}$ by the whole numbers 2, 3, 4, 5, 6, and 7.

2. Start by deciding whether the fraction is in the interval from 0 to $\frac{1}{2}$ or the interval from $\frac{1}{2}$ to 1. Then decide to which endpoint of the interval the fraction is closest. You can think of dividing the number line into the number of equal parts indicated by the denominator, and use the numerator to determine approximately where the fraction would be on the number line. This will give you an idea of which endpoint the fraction is closest to.

3. If the denominators are the same, the fraction with the greater numerator is larger. If the numerators are the same, the fraction with the smaller denominator is larger. If the denominators are different, check to see if one fraction is between 0 and $\frac{1}{2}$ and the other is between $\frac{1}{2}$ and 1. If this is true, the fraction between $\frac{1}{2}$ and 1 is larger. If both the numerators and the denominators are different, try to change one of the fractions to an equivalent fraction with the same denominator as the other fraction or find an equivalent fraction for each fraction so that the denominators are the same. Then, compare the numerators.

In this investigation, you explored equivalent fractions and compared fractions to benchmarks and other fractions. These questions will help you summarize what you have learned:

1 Find six fractions that are equivalent to $\frac{3}{5}$. Explain how you found the fractions.

2 How can you decide whether a given fraction is closest to 0, $\frac{1}{2}$, or 1?

3 How can you compare any two fractions to decide which is largest?

Think about your answers to these questions, discuss your ideas with other students and your teacher, and then write a summary of your findings in your journal.

TEACHING THE INVESTIGATION

2.1 • Comparing Notes

Launch

Tell the class the story of the teachers' debate about how best to report their fund-raising progress.

Explore

Problem 2.1 can be done rather quickly using a think-pair-share strategy. First, have students *think* about the problem on their own for 3 to 5 minutes. Then put the students in *pairs,* and have them *share* their ideas and try to reach consensus. If the students in a pair disagree, have each student listen to the other explain his or her argument. If, after hearing each other's arguments, the students still do not agree, have them write separate arguments to support their positions.

Take note of students who are reasoning in creative or interesting ways and call on them during the summary. Also, watch for naive reasoning that should be explored during the summary.

Summarize

Have several pairs present their answers. If possible, make sure there is at least one presentation supporting each teacher. It is possible that no pair will think Ms. Christos is correct. If so, ask your students to explain why Ms. Christos is incorrect. Students should see that both Mr. Park and Ms. Mendoza are correct and should be able to begin an explanation of why three fourths and six eighths describe the same part of the whole.

If students understand the issues in the problem, move quickly to Problem 2.2.

2.2 • Finding Equivalent Fractions

Launch

Remind students what they discovered in Problem 2.1.

> In Problem 2.1, we found that $\frac{3}{4}$ and $\frac{6}{8}$ represent the same quantity. Can you think of another name for $\frac{3}{4}$?
>
> What about $\frac{2}{3}$? Do you know a different name for this fraction? Can you find another name by using your fraction strips?

Help students to understand the question posed in Problem 2.2. Point out that strips shown are a different length that those used in previous problems. Students need to develop flexibility in thinking about fractions modeled with strips of different lengths. At the same time, they need to recognize that comparing and interpreting fractions depend on the whole. For example, to use fraction strips to find different names for $\frac{3}{4}$, we need to look at a set of strips of a *single* length.

When you feel students understand what is being asked, have them work in small groups to explore the problem and the follow-up.

Explore

This problem is designed to encourage students to notice a pattern in finding equivalent fractions. This pattern can be generalized by the observation that you "multiply the numerator and denominator of the fraction by the same number" to produce an equivalent fraction. The *why* behind this observation is essential. *Do not give them the rule;* allow them to discover the pattern for themselves. If they do not see the pattern, wait until the summary to ask questions that will focus them and help them to see the pattern.

Circulate as students work, asking questions about why they think they are correct and how they are finding equivalent fractions. Students should use both mathematical and informal language to explain their thinking. They need to discuss their reasoning. Try not to assist them by filling in the missing parts of their explanations; just continue to guide them through your questions.

Summarize

Have a few students report on the patterns they discovered that were helpful for finding equivalent fractions. Many students will see that they can generate equivalent fractions by multiplying the numerator and the denominator by the same number. Some students may see the pattern as multiplying by 2 repeatedly, which is actually only part of the pattern. For example, in generating fractions equivalent to $\frac{3}{4}$, they would find $\frac{3}{4}$, $\frac{6}{8}$, $\frac{12}{16}$, and so on, skipping the other equivalent fractions. Ask questions to expose other possibilities, such as $\frac{9}{12}$ and $\frac{15}{20}$. You want students to propose the idea that you can multiply the numerator and the denominator of the original representation of the fraction by the same number—any whole number—to obtain an equivalent fraction. Once this idea comes up, be sure to ask:

Why do you you think this works?

It is easy to get students to routinely multiply the numerator and denominator of a fraction by the same number to produce an equivalent fraction. However, it is important for them to build visual images and language to help them understand why this method works. They will need to spend time drawing pictures and comparing strips to really understand the concept. Use the following example to help explain the idea.

Let's use fraction strips to explore what happens when we multiply the numerator and the denominator of a fraction by the same number.

I can represent the fraction $\frac{1}{3}$ using a fraction strip. The denominator tells me to divide the whole strip into three equal parts. The numerator tells me that I am concerned with one of these three parts. (*Draw the following picture on the board, or show the first diagram on Transparency 2.2B.*)

What does it mean when I multiply the denominator of a fraction by a number? For example, what happens if I multiply the denominator of $\frac{1}{3}$ by 5?

Multiplying the denominator by 5 gives a new denominator of 15, so the strip is now divided into 15 equal parts. This means that each of the 3 parts of the original strip is now divided into 5 equal parts. Divide each section of your drawing into 5 equal parts, or show the second diagram on the transparency.

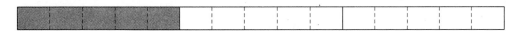

What does it mean when I multiply the numerator by the same number I multiplied the denominator by? For example, what happens when I multiply the numerator of $\frac{1}{3}$ by 5?

The new numerator, 5, tells you that you are concerned with 5 of these 15 parts. This make sense because, since each part of the original fraction strip has been divided into five pieces, it takes five times as many pieces to represent the same quantity. You can see from the drawing that 5 of these 15 smaller parts is exactly the same as 1 of the 3 original parts. That is, $\frac{5}{15}$ is equivalent to $\frac{1}{3}$.

$$\frac{1}{3} = \frac{5}{15}$$

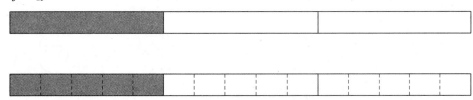

Help students understand that when they multiply the numerator and the denominator by the same number, they are actually multiplying by 1.

When I multiply the numerator and the denominator by the same number—for example, by 3—it is the same as multiplying by $\frac{3}{3}$. What is another name for this fraction?

Some students may suggest $\frac{4}{4}$ or $\frac{6}{6}$, but help them see that all of these fractions are equivalent to 1. Since the fraction is being multiplied by 1, the resulting fraction must have the same value as the original fraction.

2.3 • Making a Number Line

Launch

In this problem, students build a number line from 0 to 1, using information from all of their fraction strips. This problem enhances their understanding of equivalent fractions.

Launch the problem by demonstrating how to transfer the marks and labels from the halves and the thirds fraction strips to a number line. This process is described in the student edition.

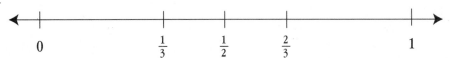

You may want to demonstrate using the fourths strip as well, so that the issue of how to label numbers with more than one name arises.

Explore

Circulate as students work individually, asking questions to keep them focused on why fractions they used to label the same mark are equivalent. As student finish, have them work on the follow-up, which asks them to add labels for fractions not included on their fraction strips.

Summarize

Display a few examples of student work, and talk about the equivalent fractions that are visible on the number lines.

This is a good time to revisit the meaning of fractions and to point toward finding equivalent fractions. Here are some questions you might ask:

> Which fraction strip shows a fraction equivalent to $\frac{2}{3}$ and a fraction equivalent to $\frac{3}{4}$?

Students can find this answer by inspecting $\frac{2}{3}$ and $\frac{3}{4}$ on their number lines. The marks for both of these numbers are also named by fractions with denominators of 12. The mark for $\frac{2}{3}$ is also labeled $\frac{8}{12}$, and the mark for $\frac{3}{4}$ is also labeled $\frac{9}{12}$. This indicates that the twelfths fraction strip shows fractions equivalent to both $\frac{2}{3}$ and $\frac{3}{4}$.

> What fraction strip could we add to our set that would also show a fraction equivalent to $\frac{2}{3}$ and a fraction equivalent to $\frac{3}{4}$?

Here you hope students begin to make the connection to multiples by observing, for example, that a twenty-fourths strip would show a name for both fractions.

> How can finding fractions equivalent to $\frac{2}{3}$ and $\frac{3}{4}$ that have the same denominator help you to compare $\frac{2}{3}$ and $\frac{3}{4}$?

When two fractions have the same denominator, you can compare their numerators; the fraction with the greater numerator is larger. Since $\frac{9}{12}$ is larger than $\frac{8}{12}$, $\frac{3}{4}$ is larger than $\frac{2}{3}$.

> Which is larger, $\frac{4}{6}$ or $\frac{7}{10}$? How do you know?

Students can find these fractions on their strips or on the number line and make visual comparisons. They can, but are not likely to, find equivalent fractions and compare them. Since 30 is a common multiple of 6 and 10, $\frac{4}{6}$ can be written $\frac{20}{30}$ and $\frac{7}{10}$ can be written $\frac{21}{30}$. Now it is clear that $\frac{7}{10}$ is larger—in fact, exactly $\frac{1}{30}$ larger.

2.4 • Comparing Fractions to Benchmarks

Launch

In this problem, students compare fractions to benchmarks. Benchmarks are numbers that are easy referents. For work with fractions, good benchmarks are 0, $\frac{1}{2}$, and 1.

Discuss the material on page 23 of the student edition. The text suggests that one way to estimate the size of a fraction is to compare it to 0, $\frac{1}{2}$, and 1. An example can help to make the method clear. You might use $\frac{2}{5}$, which is not in the list given in the problem.

> Which two benchmarks is $\frac{2}{5}$ between? Which benchmark is it closest to?

Some students may reason that if you divide a number line from 0 to 1 into fifths, it takes two and a half of these fifths to make one half. So, $\frac{2}{5}$ must be between 0 and $\frac{1}{2}$. Since two fifths is closer to two and a half fifths than it is to zero fifths, $\frac{2}{5}$ is closer to $\frac{1}{2}$ than to 0.

When you feel students understand this kind of reasoning, let them work in small groups on the problem.

Explore

Circulate as the groups work, asking students to explain their strategies for determining which benchmark each fraction is near and for ordering the fractions.

Summarize

Have students share strategies for making comparisons to benchmarks and for ordering fractions. Collect students' answers to part A and record the results in two lists on the board.

Between 0 and $\frac{1}{2}$	Between $\frac{1}{2}$ and 1
$\frac{1}{5}$	$\frac{2}{3}$
$\frac{3}{12}$	$\frac{8}{10}$
$\frac{3}{8}$	$\frac{3}{5}$
$\frac{2}{9}$	$\frac{5}{6}$
$\frac{1}{3}$	$\frac{5}{8}$
	$\frac{4}{5}$
	$\frac{3}{4}$
	$\frac{7}{12}$

> Look at the information displayed in the two lists. What do you notice about all the fractions in the interval from 0 to $\frac{1}{2}$?

Students usually notice that all the fractions have a numerator that is less than half its denominator.

What do you notice about all the fractions in the interval from $\frac{1}{2}$ to 1?

The numerator is more than half of the denominator for each of these fractions.

You may want to continue by collecting students' answers to part B and displaying the results in a table.

Closest to 0	Halfway between 0 and $\frac{1}{2}$	Closest to $\frac{1}{2}$	Halfway between $\frac{1}{2}$ and 1	Closest to 1
$\frac{1}{5}$	$\frac{3}{12}$	$\frac{3}{8}$	$\frac{3}{4}$	$\frac{8}{10}$
$\frac{2}{9}$		$\frac{1}{3}$		$\frac{5}{6}$
		$\frac{2}{3}$		$\frac{4}{5}$
		$\frac{3}{5}$		
		$\frac{5}{8}$		
		$\frac{7}{12}$		

Ask students to explain why $\frac{3}{8}$ is closer to $\frac{1}{2}$ than to 0 and why $\frac{3}{5}$ is closer to $\frac{1}{2}$ than to 1. Spend some time discussing where $\frac{3}{12}$ and $\frac{3}{4}$ belong. Students often want to "round up" and say that these numbers are closer to the right-hand benchmark. Take some time to discuss why rounding is not the issue in this problem. The question is concerned with the benchmarks these numbers are closest to in terms of *distance*, and these numbers are located exactly between two benchmarks.

To summarize part D, you may want to have students place the fractions on a large number line marked with 0, $\frac{1}{2}$, and 1. Ask students how their work with benchmarks in parts A and B helped them to order the fractions and what additional work they had to do to complete part D.

Work as a class on the follow-up. These questions provide an opportunity to review the methods the class has developed so far for comparing two fractions. Discussing these questions will help students review and synthesize what they know *right now*. We suggest that you work as a class to build a list of strategies for comparing fractions. Make sure students can explain why each strategy works. Here are some ideas the list might include:

- If the denominators are the same, compare the numerators. The fraction with the greater numerator is larger.

- If the numerators are the same, compare the denominators. The fraction with the smaller denominator is larger.

- If the denominators are different, check whether one fraction is between 0 and $\frac{1}{2}$ and the other is between $\frac{1}{2}$ and 1. If this is true, the fraction between $\frac{1}{2}$ and 1 is larger.

- If both the numerators and the denominators are different, try to change one of the fractions to an equivalent fraction with the same denominator as the other fraction. Then, compare the numerators.

- If both the numerators and the denominators are different, find equivalent fractions for both fractions so that the denominators are the same. Then, compare the numerators.

Discuss the fact that the fewer "rules" students must remember, the better off they will be. Ask students for ideas about how they could have fewer rules and still be able to make the necessary comparisons. Don't insist on formalizing the algorithm (that is, to find equivalent fractions with like denominators). Rather, help students go as far as they can in generalizing a strategy for finding equivalent fractions.

Continue working with your students, offering examples of fractions to compare or having them make up examples for each other. You may want to create a deck of index cards showing a variety of possible fractions, such as:

$$\frac{1}{2} \quad \frac{2}{2} \quad \frac{3}{2} \quad \frac{4}{2} \quad \frac{5}{2} \quad \frac{6}{2}$$

$$\frac{1}{3} \quad \frac{2}{3} \quad \frac{3}{3} \quad \frac{4}{3} \quad \frac{5}{3} \quad \frac{6}{3}$$

$$\frac{1}{4} \quad \frac{2}{4} \quad \frac{3}{4} \quad \frac{4}{4} \quad \frac{5}{4} \quad \frac{6}{4} \quad \frac{7}{4} \quad \frac{8}{4}$$

$$\frac{1}{5} \quad \frac{2}{5} \quad \frac{3}{5} \quad \frac{4}{5} \quad \frac{5}{5} \quad \frac{6}{5} \quad \frac{7}{5} \quad \frac{8}{5} \quad \frac{9}{5} \quad \frac{10}{5}$$

$$\frac{1}{6} \quad \frac{2}{6} \quad \frac{3}{6} \quad \frac{4}{6} \quad \frac{5}{6} \quad \frac{6}{6} \quad \frac{7}{6} \quad \frac{8}{6} \quad \frac{9}{6} \quad \frac{10}{6}$$

$$\frac{1}{8} \quad \frac{2}{8} \quad \frac{3}{8} \quad \frac{4}{8} \quad \frac{5}{8} \quad \frac{6}{8} \quad \frac{7}{8} \quad \frac{8}{8} \quad \frac{9}{8} \quad \frac{10}{8}$$

$$\frac{1}{10} \quad \frac{2}{10} \quad \frac{3}{10} \quad \frac{4}{10} \quad \frac{5}{10} \quad \frac{6}{10} \quad \frac{7}{10} \quad \frac{8}{10} \quad \frac{9}{10} \quad \frac{10}{10} \quad \frac{11}{10}$$

$$\frac{1}{12} \quad \frac{2}{12} \quad \frac{3}{12} \quad \frac{4}{12} \quad \frac{5}{12} \quad \frac{6}{12} \quad \frac{7}{12} \quad \frac{8}{12} \quad \frac{9}{12} \quad \frac{10}{12} \quad \frac{11}{12} \quad \frac{12}{12}$$

Students can turn two index cards face up and compare the two fractions, recording the fractions and the result of their comparison. You might challenge them to arrange all the cards in pairs so that the first fraction in each pair is larger than the second fraction.

Periodically, you can return to the class list of strategies to see whether students have found better ways to make comparisons.

2.5 • Fractions Greater Than One

Launch

Making the transition from fractions between 0 and 1 to fractions greater than 1 is often difficult for students. This problem is designed to help them see that they can reason about fractions within the interval between any two consecutive whole numbers in the same way they reason about fractions between 0 and 1. To write a fraction greater than 1 as a mixed number, students can find the fraction portion the same way they find fractions between 0 and 1; the whole number portion is simply the whole number on the number line immediately to the left of the fraction.

This is a good time to begin making a class number line that extends from 0 to 10 (or beyond) with consecutive whole numbers exactly 16 cm apart (so that the class can use their fraction strips to fill in the in-between points).

So far, we have worked with the part of the number line between 0 and 1. In this problem, we will look at the part of the number line between 1 and 2.

Show a 16 cm number line from 1 to 2 on the overhead projector, or demonstrate using the section of the class number line from 1 to 2.

> How can we use our strips to find fractions between these two whole numbers? (*Ask a student who has an idea to demonstrate.*)
>
> How would we label this amount? (*Point to the halfway mark.*) How about this amount? (*Point to the mark that is one fourth of the way from 1 to 2.*)

Discuss how to label the point halfway between 1 and 2 as both a mixed number $(1\frac{1}{2})$ and an improper fraction $(\frac{3}{2})$. You may need to ask how many halves are in $1\frac{1}{2}$ to get students to think about labeling points with improper fractions. Then, ask them to help you label the point one fourth of the way between 1 and 2 with both an improper fraction and a mixed number. This point can be labeled $1\frac{1}{4}$. Since there are four fourths in 1, plus another fourth after the 1, it can also be labeled $\frac{5}{4}$.

> What are some ways we can reason about how to label marks on this number line without using our fraction strips?

Students may suggest making a copy of the number line and folding it. Others might want to measure the line and find what part of the length each mark represents. Since the number line is 16 cm in length, this strategy can be used to label some of the points quite easily. For instance, the fourths marks will be 4 cm apart.

> With your partner, look at the number line on page 24. Use your strips, or other strategies that make sense to you, to label all of the lettered points. Add to your answers any equivalent labels for the flagged points that have denominators of 12 or smaller. Label the points with both mixed numbers and improper fractions.

Explore

Some pairs will need help labeling the points with improper fractions. Help these pairs by asking them questions.

> How many parts of this size would be in the whole? We have this many more parts of this size. How many does that make in all?

When students seem ready, explain part B of Problem 2.5.

> The second part of the problem asks you to add and label some points. The problem has requirements on where the points are to be added. Read these carefully, and work with your partner to find good answers that you can defend.

If pairs finish early, ask them to think about the follow-up.

Summarize

Have students share their answers and strategies for labeling the lettered points. You could choose a pair of students to label the marks representing the lettered points on the class number line. Challenge other pairs to fill other intervals on the number line. You might assign a pair of students to each whole number interval. Leave this number line on display; adding to it is a good way for students to continue to build an understanding of fractions as numbers that measure lengths in between whole numbers.

If you think your students are ready, pose the following questions to encourage them to think about an efficient way to convert a mixed number to an improper fraction and vice versa.

> I noticed an interesting pattern when you were presenting your evidence on this problem. Look at this mixed number. (*Write $4\frac{3}{5}$ on the board.*) How can we write this number as an improper fraction—that is, as a fraction whose numerator is larger than its denominator?

Students should be able to reason that since there are 20 fifths in 4, plus 3 more fifths in $\frac{3}{5}$, the improper fraction representation is $\frac{23}{5}$.

> (*Point to the appropriate numbers in the $4\frac{3}{5}$ you have written on the board as you explain the following.*) If I multiply 4 by 5 and add 3, I get 20 + 3, or 23. (*Write 20 + 3 = 23 on the board.*) This is the *numerator* of the improper fraction representation. Do you suppose this will always work? Why?

This always works. When we multiply the whole number, 4, by the denominator, 5, the result tells us how many fifths there are in the whole number. When we add 3, we are finding the total number of fifths in the mixed number. The result, 23, becomes the numerator, and 5 becomes the denominator.

> I wonder whether we can go the other way. How would we write a number like $\frac{11}{3}$ as a number with a whole number part and a fraction less than 1?

Ask students to share their ideas. Hopefully they will see that they need to make as many wholes as possible and write the parts left over as a fraction. In this example, we have 11 parts. It takes 3 parts to make one whole. There are 3 groups of 3 parts in 11, with 2 parts left over. This gives us 3 wholes and 2 thirds, or $3\frac{2}{3}$. If your students do not come up with this, just continue to ask questions as opportunities arise during the remainder of the unit. The terms *improper fraction* and *mixed number* are not important; they are just convenient ways to be more precise. Your students may do just fine using their own informal words to describe these two kinds of representation.

Additional Answers

Answers to Problem 2.3

A.

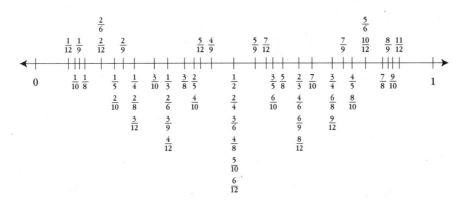

Answers to Problem 2.4 Follow-Up

1. Explanations will vary.
   a. $\frac{3}{12} < \frac{7}{12}$
   b. $\frac{5}{6} > \frac{5}{8}$
   c. $\frac{2}{3} > \frac{3}{9}$
   d. $\frac{13}{12} < \frac{6}{5}$

2. Explanations will vary.
   a. $\frac{4}{7} < \frac{6}{7}$
   b. $\frac{7}{10} > \frac{8}{12}$
   c. $\frac{5}{8} < \frac{6}{9}$
   d. $\frac{10}{12} = \frac{5}{6}$
   e. $\frac{2}{4} < \frac{5}{9}$
   f. $\frac{3}{9} > \frac{3}{10}$

ACE Questions

Applications

6.

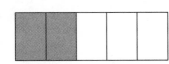

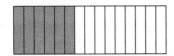

7.

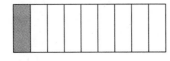

Connections

22.

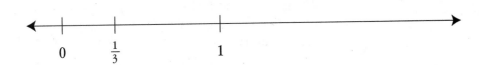

23.

37a.

300 km

180 km

240 km

Cooking with Fractions

Mathematical and Problem-Solving Goals

- **To continue building an understanding of equivalent fractions**

- **To explore the use of squares and other areas as a way to build visual models of fractional parts of a whole**

- **To explore real-life problems that require operations on fractions in a context that invites the use of informal strategies rather than formal rules and algorithms**

So far in this unit, we have been using fractions to express parts of a whole. Our visual model for representing fractions has been the fraction strip, which is related to a number line and a ruler. In this investigation, students are introduced to another powerful visual representation of fractions: representations that model fractions as subdivisions of areas of figures.

In one sense, the fraction strip is an area model. However, we have ignored width and focused on length as the salient feature of the model. In this investigation, we focus on ways to subdivide the area of a figure to model fractions. While rectangles and squares are nice figures to work with because they are easy to subdivide, we include some nonrectangles in the ACE questions to help students see that, theoretically, we can use any area as long as we can divide it into equal-size parts.

In Problem 3.1, Area Models for Fractions, students explore the possible ways to cut a square pan of brownies into 15 equal-size large brownies, 20 equal-size medium brownies, and 30 equal-size small brownies. Problem 3.2, Baking Brownies, challenges students to adjust a recipe to make enough brownies to serve a given number of students.

Student Pages	**31–38**
Teaching the Investigation	**38a–38g**

Materials

For students

- Labsheet 3.1 (1 per student)
- Rulers or other straightedges

For the teacher

- Transparencies 3.1, 3.2A, and 3.2B (optional)
- Transparencies of Labsheet 3.1 (optional)

INVESTIGATION 3

Cooking
with Fractions

You have made and used fraction strips to help you think about fractions as parts of wholes. You have used your strips to name fractional amounts, to compare fractions, to find equivalent fractions, and to make a number line showing fractions between whole numbers.

3.1 Area Models for Fractions

You can also think about fractions as parts of a region. For example, if a pizza is cut into eight slices of the same size, and you eat two of the slices, you have eaten two eighths of the pizza. If you eat five of the slices, you have eaten five eighths of the pizza.

$\frac{2}{8}$ **has been eaten**

$\frac{5}{8}$ **has been eaten**

If you divide a square pan of brownies into equal-size brownies and then eat two brownies, what part of the batch have you eaten? To answer this question, you need to know the total number of brownies in the batch.

Problem 3.1

Use the squares on Labsheet 3.1 as models for pans of brownies. Show the cuts you would make to divide a pan of brownies into

A.   15 equal-size large brownies

B.   20 equal-size medium brownies

C.   30 equal-size small brownies

Investigation 3: Cooking with Fractions   **31**

Answers to Problem 3.1

See page 38d.

Answers to Problem 3.1 Follow-Up

1. $\frac{1}{30}$, $\frac{1}{20}$, $\frac{1}{15}$; If a pan of brownies is divided into 30 equal-size pieces, then one of these pieces is $\frac{1}{30}$ of the whole pan. Explanations for the medium and large brownies are similar.

2.   a.   There are two ways; see answer to Problem 3.1, part A.
     b.   There are three ways; see answer to Problem 3.1, part B.
     c.   There are four ways; see answer to Problem 3.1, part C.

Launch

■ Elicit ideas about how a pan of brownies could be cut into equal-size pieces.

Explore

■ As students work, remind them of things they already know that may help.

■ Encourage students to connect the brownie-cutting problem to factors of numbers.

■ Have groups that are thinking in creative ways record their answers on transparencies of Labsheet 3.1.

Summarize

■ Have students explain their results.

■ Ask questions to help students think more deeply about the problem.

■ Have students create their own brownie-cutting problems. (*optional*)

Assignment Choices

ACE questions 13, 18–23, 24–32, and unassigned choices from earlier problems

Baking Brownies

Launch

- Read the story of the students' brownie-making task aloud.

Explore

- If a group has trouble, ask them how many batches of brownies they would need to make to feed 240 people.

Summarize

- Have students share their ideas about the solutions.

Assignment Choices

ACE questions 14–17, and unassigned choices from earlier problems

Assessment

It is appropriate to use the quiz after this problem.

■ Problem 3.1 Follow-Up

1. What fraction of a whole pan is one small brownie? One medium brownie? One large brownie? Explain.

2. **a.** Is there more than one way to cut a pan of brownies into 15 equal-size large brownies? If so, show the other ways. If not, explain why it cannot be done.

   b. Is there more than one way to cut a pan of brownies into 20 equal-size medium brownies? If so, show the other ways. If not, explain why it cannot be done.

   c. Is there more than one way to cut a pan of brownies into 30 equal-size small brownies? If so, show the other ways. If not, explain why it cannot be done.

3.2 Baking Brownies

Next week, the eighth graders from Sturgis Middle School are attending school camp. Samantha, Romero, and Harold have the job of making brownies for an afternoon snack for the entire camp—all 240 people!

Chunky Brownies with a Crust

$1\frac{1}{4}$ cups flour

$\frac{1}{4}$ cup sugar

$\frac{1}{2}$ cup cold butter or margarine

1 14-ounce can sweetened condensed milk

$\frac{1}{4}$ cup unsweetened cocoa

1 egg

1 teaspoon vanilla

$\frac{1}{2}$ teaspoon baking powder

1 7-ounce bar milk chocolate, broken into small chunks

$\frac{3}{4}$ cup chopped nuts (optional)

Preheat the oven to 350 degrees. In a medium bowl, combine 1 cup of flour and the sugar. Cut in the margarine or butter until crumbly. Press the mixture firmly into the bottom of a 10-by-10-inch baking pan. Bake 15 minutes. Meanwhile, in a large mixing bowl, beat the sweetened condensed milk, the cocoa, the egg, the remaining flour, the vanilla, and the baking powder. Stir in the nuts and chocolate chunks. Spread over the prepared crust. Bake 20 minutes. Cool. Sprinkle with confectioner's sugar if desired. Store tightly covered at room temperature. Makes 15 large, 20 medium, or 30 small brownies.

Problem 3.2

A. Do you think Samantha, Romero, and Harold should make small, medium, or large brownies?

B. If they make brownies of the size you chose in part A, how much of each ingredient will they need to make enough to serve a brownie to each person at camp?

C. Describe the strategy you used to get your answer to part B.

■ Problem 3.2 Follow-Up

Compare your answers for part B to the answers of classmates who did calculations for the other two sizes.

Suppose you get to decide which size brownies will be served to the campers. Tell which size you would choose in each situation below. Explain your answer.

1. You are in charge of buying the ingredients, and you have a limited budget.

2. You have to help make the brownies.

3. You don't have to do any work, you just get to eat the brownies.

Answers to Problem 3.2

A. Answers will vary.

B. See the table on page 38c.

C. Answers will vary.

Answers to Problem 3.2 Follow-Up

1. Most students will choose the small brownies because they would cost the least to make.

2. Answers will vary.

3. Answers will vary.

Answers

Applications

1. Possible answer:

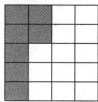

2. Possible answer:

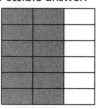

3. Possible answer:

4. Possible answer:

5. Possible answer:

6. See right.

7. See right.

8. See page 38d.

9. See page 38e.

10. $\frac{1}{8}$

11. $\frac{5}{16}$

12. $\frac{9}{28}$

13. $\frac{10}{64} = \frac{5}{32}$

Applications • Connections • Extensions

As you work on these ACE questions, use your calculator whenever you need it.

Applications

In 1–4, illustrate each fraction by drawing a square, subdividing it into equal-size regions, and shading the fractional part indicated.

1. $\frac{7}{20}$ 2. $\frac{3}{15}$ 3. $\frac{12}{18}$ 4. $\frac{3}{7}$

In 5–8, illustrate each fraction by drawing a square and subdividing it in a different way than you did for questions 1–4.

5. $\frac{7}{20}$ 6. $\frac{3}{15}$ 7. $\frac{12}{18}$ 8. $\frac{3}{7}$

9. Show $\frac{3}{12}$ in three different ways by subdividing and shading a square.

In 10–13, tell what fractional part of the whole figure is shaded.

10.

11.

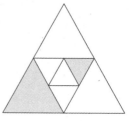

12.

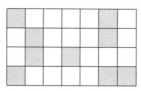

13.

6. Possible answer:

7. Possible answer:

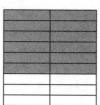

14. Gerhard wants to make Tahini Granola Cookies to bring on a hiking trip.

Tahini Granola Cookies

These cookies are an excellent high-energy snack
to take on camping trips or hikes.

$\frac{1}{2}$ cup tahini $\qquad$ $\frac{1}{2}$ cup melted butter
$\frac{2}{3}$ cups honey $\qquad$ 2 cups of your favorite granola
2 teaspoons vanilla $\qquad$ $\frac{1}{4}$ cup chopped nuts

Preheat the oven to 350°. Blend the tahini and honey together. Add the vanilla
and butter and mix well. Add the granola and nuts and mix well. Drop by
tablespoon, $1\frac{1}{2}$ to 2 inches apart, on the cookie sheet and bake 10–15 minutes
or until golden brown. Transfer to a dry surface to cool. Makes about 3 dozen
cookies.

a. How much of each ingredient will Gerhard need to make five batches of
granola cookies?

b. Instead of dropping spoonfuls of dough onto the pans, Gerhard pressed the
dough evenly into square pans. When the granola cookie mix had baked, he
cut it into bars. Show several ways Gerhard can cut a pan of granola cookies
to get three dozen bars.

Connections

15. a. How many fourths are in $4\frac{1}{4}$? (A number like $4\frac{1}{4}$ is called a *mixed number*—it
is a mix of a whole number and a fraction.)

b. Use your answer to part a to write $4\frac{1}{4}$ as a fraction with a denominator of 4.
(This is sometimes called changing the form of a number from a mixed
number to an *improper fraction*.)

14a. tahini: $\frac{5}{2}$ or $2\frac{1}{2}$ cups

honey: $\frac{10}{3}$ or $3\frac{1}{3}$ cups

vanilla: 10 teaspoons

butter: $\frac{5}{2}$ or $2\frac{1}{2}$ cups

granola: 10 cups

nuts: $\frac{5}{4}$ or $1\frac{1}{4}$ cups

14b. See page 38e.

15a. There are 17 fourths
in $4\frac{1}{4}$.

15b. $\frac{17}{4}$

Connections

16a. There are 8 fifths in $1\frac{3}{5}$.

16b. $\frac{8}{5}$

17a. There are 23 sixths in $3\frac{5}{6}$.

17b. $\frac{23}{6}$

18. See below right.

19. See page 38e.

20. See page 38e.

21. See page 38f.

22. See page 38f.

23. See page 38f.

24. $\frac{2}{5}$

25. $\frac{1}{9}$

26. Possible answer:

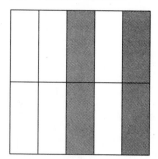

27. Possible answer:

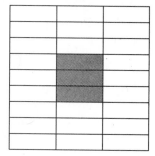

16. **a.** How many fifths are in $1\frac{3}{5}$?

b. Use your answer to part a to write $1\frac{3}{5}$ as a fraction with a denominator of 5.

17. **a.** How many sixths are in $3\frac{5}{6}$?

b. Use your answer to part a write $3\frac{5}{6}$ as a fraction with a denominator of 6.

In 18–23, draw, subdivide, and shade square regions to illustrate and complete each statement. For example, the statement $\frac{?}{10} = \frac{3}{5}$ can be illustrated like this:

 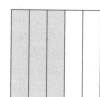

The illustration shows that $\frac{6}{10} = \frac{3}{5}$.

18. $\frac{3}{15} = \frac{?}{30}$ **19.** $\frac{18}{30} = \frac{?}{10}$ **20.** $\frac{1}{2} = \frac{?}{20}$

21. $\frac{?}{15} = \frac{3}{5}$ **22.** $\frac{?}{20} = \frac{3}{5}$ **23.** $\frac{9}{15} = \frac{?}{30}$

In 24 and 25, tell what fraction of each square is shaded.

24.   **25.**

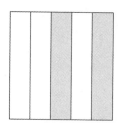   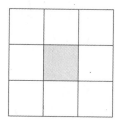

26. Draw a picture to illustrate a fraction with a denominator of 10 that is equivalent to the fraction shown in question 24.

27. Draw a picture to illustrate a fraction with a denominator of 27 that is equivalent to the fraction illustrated in question 25.

18. $\frac{3}{15} = \frac{6}{30}$. Possible illustrations:

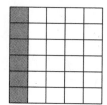 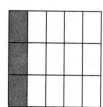

28. Order the following fractions from smallest to largest:

$$1\frac{7}{10} \qquad \frac{5}{3} \qquad 1\frac{12}{18} \qquad \frac{25}{15}$$

In 29–31, use this drawing of a portion of a ruler. The numbers indicate inches.

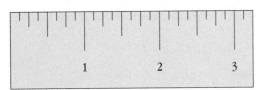

29. What fraction does each mark between the left end of the ruler and the 1-inch mark represent?

30. If the smallest sections of the ruler were each divided into two equal parts, how should the new parts between 0 and 1 be labeled?

31. What fractions do the marks between 1 inch and 2 inches represent?

Extensions

32. If a 13-by-9-inch brownie pan is divided into 20 equal-size brownies, what are the dimensions of one brownie?

28. $\frac{5}{3} = \frac{25}{15} = 1\frac{12}{18}$, $1\frac{7}{10}$ ($\frac{5}{3}$, $\frac{25}{15}$, and $1\frac{12}{18}$ are equivalent, and all are less than $1\frac{7}{10}$)

29. See below left.

30. See below left.

31. See page 38f.

Extensions

32. See page 38g.

29.

30.

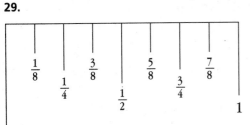

38 Investigation 3

Possible Answers

1. You look at the denominator of the fraction to see how many equal parts to divide the square into. Then you shade the number of parts given by the numerator of the fraction.

2. You can compare the square models to determine which has a larger shaded area; this model represents the larger fraction.

3. You count the total number of equal parts; this is the denominator of the fraction. Then you count the number of shaded parts; this is the numerator of the fraction.

Mathematical Reflections

In this investigation, you divided squares and rectangles into regions to help you model fractions. These questions will help you summarize what you have learned:

1 Describe your strategy for dividing a square to represent a fraction.

2 How can square models help you decide which of two fractions is larger?

3 How do you find the fraction name for a shaded part of a square? Use a specific example if it helps you to think about the process.

Think about your answers to these questions, discuss your ideas with other students and your teacher, and then write a summary of your findings in your journal.

Tips for the Linguistically Diverse Classroom

Original Rebus The Original Rebus technique is described in detail in *Getting to Know CMP*. Students make a copy of the text before it is discussed. During discussion, they generate their own rebuses for words they do not understand as the words are made comprehensible through pictures, objects, or demonstrations. Example: Question 3—key words for which students may make rebuses are *fraction name* ($\frac{a}{b}$), *shaded* (a shaded square), *square* (a square), *example* (a square with $\frac{1}{4}$ shaded).

TEACHING THE INVESTIGATION

3.1 • Area Models for Fractions

Launch

Make sure students understand the context of the problem.

> How many of you have made a pan of brownies and cut it into pieces?

Draw a square on the board or at the overhead projector.

> Suppose this square represents a pan of brownies hot out of the oven. If you wanted to cut brownies that are all the same size, how many brownies would you cut? How would you make the cuts?

Try to elicit more than one suggestion. You or students can demonstrate the ideas as they are raised.

> In this problem, you will explore the ways you can cut a pan of brownies to get 30 small brownies, 20 medium brownies, or 15 large brownies.

Give each person a copy of Labsheet 3.1, which has several squares for modeling brownie pans.

Explore

Have students work in small groups or pairs to explore the problem and the follow-up. There are several ways for students to cut the brownies to get 15 large, 20 medium, or 30 small brownies. Some groups will have a hard time finding one or two ways; others will quickly find several.

You may be surprised at how hard it is for students to draw reasonable representations of their cutting strategies. Remind them that they know folding strategies that may help. They can fold strips of paper with the same length as a square side into fractional parts and use the folded strips as guides. Some students will try to measure with a ruler. This can be difficult if the divisions of the measured length do not align with marks on the ruler.

Be on the lookout for students who are connecting this problem to factors of numbers as explored in the *Prime Time* unit. You might encourage this connection by asking questions about the factors of 15, 20, and 30. The rectangles students made for these numbers in *Prime Time* can help them here. Note that in this problem the factor pairs do not represent the dimensions of rectangles, but the number of strips cut in each direction to produce the given number of pieces.

Look for groups that are thinking about the problem in an organized, interesting, or creative way. Give these groups transparencies of Labsheet 3.1 on which to record their answers. They can use these transparencies to present their strategies during the summary.

Summarize

Have students share their answers and strategies. Ask them to explain why the cuts they made will work.

What part of the whole pan would a small brownie be? ($\frac{1}{30}$) What part of the whole pan would a medium brownie be? ($\frac{1}{20}$) A large brownie? ($\frac{1}{15}$).

How would you describe the shape of a small brownie cut in each of the ways we have found? (*They range from long and very skinny to almost square.*)

Are all the small brownies the same size (area) even though they have different shapes? (*yes; They are all $\frac{1}{30}$ of a pan.*)

If I ate three small brownies, two medium brownies, and one large brownie, how much of a pan of brownies would I have eaten? ($\frac{4}{15}$, or $\frac{8}{30}$, or $\frac{16}{60}$) How would I feel?

Here students must add $\frac{3}{30} + \frac{2}{20} + \frac{1}{15}$. Let them try to devise strategies for tackling the problem. If the strategy of finding a common multiple and renaming the fractions does not come up, suggest it yourself.

Would it be easier to find the total amount if we renamed the fractions so they all had the same denominator? Why would this be a helpful strategy? What denominator could we use?

Don't compel students to consider an algorithm for adding fractions at this time. There will be many opportunities for them to think about how to combine fractions. They will have a deeper understanding if you let them make sense of these ideas. Just continue to ask questions whenever the opportunity arises.

Always try to ask questions that reverse the information students know and the information they must figure out. In this problem, students started out knowing the number of brownies, and they determined what fraction of the whole that number of brownies represented. Here is a problem that begins with fractions.

Rodrigo ate the equivalent of a fourth of a pan of brownies before he turned green. He ate at least one brownie of each size. What are some possibilities for the number of brownies of each size that Rodrigo ate? (*The possibilities are three medium, one large, and one small brownie; or two large, one medium, and two small brownies.*)

At this point, you may want to have students make up some problems of their own about brownies of different sizes.

3.2 • Baking Brownies

Launch

Read the story about the students making brownies aloud to your class. Then have students read the problem. Make sure they understand what is being asked.

As it is written, the problem asks students to choose which size brownies to serve and then to figure out how much of each ingredient is needed. To make checking the answers more manageable, you may wish instead to have all the groups do calculations for small brownies.

Explore

Since the students do not have any formal way to add or multiply fractions, this problem is designed to encourage them to construct their own ways of combining fractions. Some of your students will exhibit some very good thinking; some will struggle. Since the fractions are simple, involving halves and fourths, most students should be able to reason through the problem.

If students are having trouble, you can get them started by asking questions.

> What size brownie did you choose? How many brownies of that size are in one batch? How many batches would you need to make 240 brownies?

One batch is 15 large, 20 medium, or 30 small brownies. Since the brownies must feed 240 people, 16 batches are needed to make large brownies, 12 batches are needed to make medium brownies, and 8 batches are needed to make small brownies.

Once they have determined the number of batches needed, students must find the amount of each ingredient required. For example, the quantity of chopped nuts required to make the large brownies is $16 \times \frac{3}{4}$. One group found this quantity by first combining $\frac{3}{4} + \frac{3}{4} + \frac{3}{4} + \frac{3}{4}$ to get $\frac{12}{4}$. They recognized this as 3 cups and then reasoned that they needed 4×3 cups or 12 cups of nuts.

Summarize

Have students share their ideas about the solutions. Here is a summary chart for each brownie size (this chart appears on Transparency 3.2B).

	Small brownie (30)	Medium brownie (20)	Large brownie (15)
Batches for 240	8 batches	12 batches	16 batches
Cups flour	10	15	20
Cups sugar	2	3	4
Cups butter	4	6	8
Cans milk	8	12	16
Cups cocoa	2	3	4
Eggs	8	12	16
Tsp. vanilla	8	12	16
Tsp. baking powder	4	6	8
No. 7-oz choc. bars	8	12	16
Cups nuts	6	9	12

Some of your students will probably see that they can obtain the values for the "Medium brownie" column by taking one and a half of the corresponding ingredient in the "Small brownie" column. Also, large brownies will take twice as many pans and twice as much of each ingredient as small brownies.

Additional Answers

Answers to Problem 3.1

A. Possible answers:

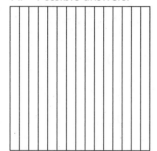 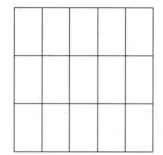

B. Possible answers:

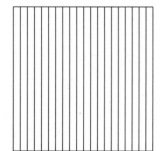 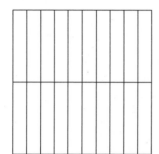 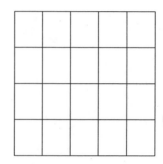

C. Possible answers:

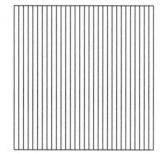 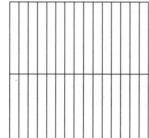 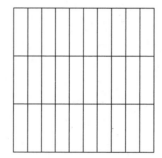 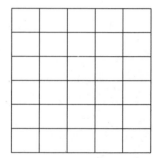

ACE Questions

Applications

8. Possible answer:

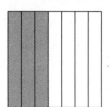

9. Possible answer:

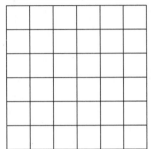 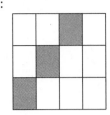 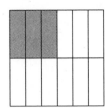

14b. Possible answers:

 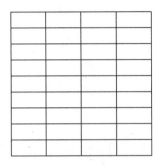 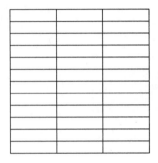

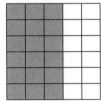

Connections

19. $\frac{18}{30} = \frac{9}{15}$. Possible illustrations:

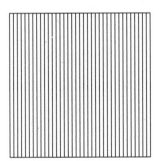

20. $\frac{1}{2} = \frac{10}{20}$. Possible illustrations:

 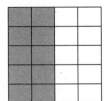

21. $\frac{9}{15} = \frac{3}{5}$. Possible illustrations:

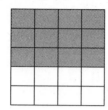

22. $\frac{12}{20} = \frac{3}{5}$. Possible illustrations:

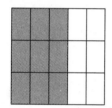 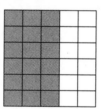

23. $\frac{9}{15} = \frac{18}{30}$. Possible illustrations:

31.

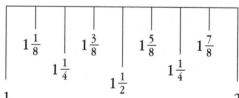

$1\frac{1}{8}$ $1\frac{3}{8}$ $1\frac{5}{8}$ $1\frac{7}{8}$

$1\frac{1}{4}$ $1\frac{1}{2}$ $1\frac{1}{4}$

1   2

Extensions

32. There are two possible ways to cut the brownies. For the first pan, each brownie has dimensions $2\frac{1}{4}" \times 2\frac{3}{5}"$ (or $\frac{9}{4}" \times \frac{13}{5}"$). For the second pan, each brownie has dimensions $1\frac{4}{5}" \times 3\frac{1}{4}"$ (or $\frac{9}{5}" \times \frac{13}{4}"$).

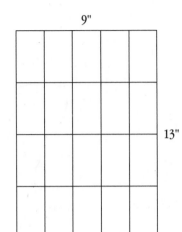

From Fractions to Decimals

Mathematical and Problem-Solving Goals

- **To extend knowledge of place value of whole numbers to decimal numbers**

- **To represent fractions with denominators of ten and powers of 10 as decimal numbers**

- **To visualize the representation of decimal numbers using a 10-by-10-grid area model**

- **To relate fraction benchmarks to decimal benchmarks**

- **To write, compare, and order decimals with place values to ten thousandths**

This investigation uses square grids as a context for introducing decimal numbers. It is assumed that students have dealt with decimals in previous grades.

In Problem 4.1, Designing a Garden, students plan a 100-square-meter garden plot, arranging it to accommodate specified vegetables. In doing so, they explore representing fractional parts of a whole. In Problem 4.2, Making Smaller Parts, students are encouraged to visualize what happens as a tenths grid is partitioned into increasingly smaller subdivisions, resulting in first a hundredths grid, then a thousandths grid, and finally a ten-thousandths grid. This promotes a sense of pattern as students think about what would be the next decimal place and how this decimal place can be represented graphically. In Problem 4.3, Using Decimal Benchmarks, benchmarks are revisited to relate fractions and decimals. In Problem 4.4, Playing Distinguishing Digits, students solve puzzles that help further their understanding of place value. The puzzles also provide opportunities for students to reason about digits using clues that connect to their work in *Prime Time.*

Materials

For students

- Labsheets 4.1, 4.2, and 4.ACE (1 of each per student)

- Transparency 4.2E and transparency markers (optional; for sharing answers with the class)

- Colored cubes or tiles (optional; 100 per group)

- Grid paper (provided as a blackline master)

- Distinguishing Digits cards (provided as blackline masters. Copy the cards, cut them out, and put them in envelopes marked with the puzzle number.)

For the teacher

- Transparencies 4.1, 4.2A–G, 4.3, and 4.4 (optional)

INVESTIGATION 4

From Fractions to Decimals

So far in this unit, you have expressed numbers as fractions. In the elementary grades, you studied another way to represent numbers—as *decimals.* Decimals are a convenient way to express fractions with denominators such as 10, 100, 1000, or 10,000. In this investigation, you will have a chance to review decimals and to connect the decimal representation of a number to the fraction representation of the same number.

 Designing a Garden

In Dayton, Ohio, the town council and the Benjamin Wegerzyn Garden Center created the largest community garden in the world. A community garden is a garden that is shared by several people. The community garden in Dayton has 1173 square plots of land that can be used by that many individual people or families to make gardens. Each plot of land has an area of 100 square meters.

Justin's family has a plot in the community garden. His father wants Justin to design the vegetable garden for his family. Justin may decide how much of the land to allocate for each type of vegetable his family wants to grow, but he must satisfy a set of conditions they put on the garden. Justin has to present the plan to his family with a drawing of the garden that specifies what fraction of the plot will be planted with each kind of vegetable.

Tips for the Linguistically Diverse Classroom

Rebus Scenario The Rebus Scenario technique is described in detail in *Getting to Know CMP*. This technique involves sketching rebuses on the chalkboard that correspond to key words in the story or information you present orally. Example: some key words and phrases for which you may need to draw rebuses while discussing the material on this page: *Dayton, Ohio* (an outline of the U.S. with an X approximately where Ohio is), *garden* (a vegetable patch), *type of vegetable* (carrots, string beans, corn, and so on), *plan* (a grid with some squares dotted and some squares striped and a key that says dots = carrots, stripes = corn).

At a Glance

Launch

■ Introduce the story of the family garden plot, making sure students understand how the hundredths grid relates to the problem.

Explore

■ Circulate while students work, asking questions to refocus students who are struggling and to further challenge groups that finish early.

Summarize

■ Have a few groups present and explain their plans.

■ Challenge the class to make designs that extend the ideas in the follow-up.

Assignment Choices

ACE questions 38–41 and unassigned choices from earlier problems

To help plan, Justin first draws a grid with 100 squares, each representing 1 square meter of the 100-square-meter plot.

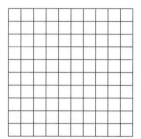

Problem 4.1

Here are the family's requirements for the garden.
- Justin's father wants to be sure potatoes, beans, corn, and tomatoes are planted. He wants twice as much of the garden to be planted in corn as potatoes. He wants three times as much land planted in potatoes as tomatoes.
- Justin's sister wants cucumbers in the garden.
- Justin's brother wants carrots in the garden.
- Justin's mother wants eggplant in the garden.
- Justin wants radishes in the garden.

Use Labsheet 4.1 to make a suitable plan for the garden. Write a description of the garden you plan. Name the fraction of the garden space that will be allotted to each kind of vegetable as part of your description. Explain how your garden will satisfy each member of Justin's family.

Problem 4.1 Follow-Up

1. Justin's father says that he will not plant less than a square meter of any vegetable. Design a garden with the largest possible amount of land planted in potatoes that fits the conditions of the problem and has at least one square meter allotted for each vegetable.

2. Design a garden with the smallest possible amount of land planted in potatoes that fits the conditions of the problem and has at least one square meter allotted for each vegetable.

Answer to Problem 4.1

Even though the relationship of square meters for vegetables chosen by Justin's father will be the same for all groups, the way the vegetables are arranged on the grid and the number of square meters given to corn, potatoes, and tomatoes will vary, as will the choices for the other vegetables. Students should have included an explanation for their design and labeled the fractional parts.

Answers to Problem 4.1 Follow-Up

1. Potatoes should make up $\frac{27}{100}$ of the garden, tomatoes $\frac{9}{100}$, and corn $\frac{54}{100}$. The remaining vegetables should make up a total of $\frac{10}{100}$ of the garden.

2. Potatoes should make up $\frac{3}{100}$ of the garden, tomatoes $\frac{1}{100}$, and corn $\frac{6}{100}$. The remaining vegetable should make up a total of $\frac{90}{100}$ of the garden.

4.2 Making Smaller Parts

Decimals give us a way to write special fractions that have denominators like 10, 100, 1000, and 10,000. A tenths grid can help you to understand decimals.

A *tenths grid* is divided into ten equal parts. It resembles the tenths fraction strip you have been using, only it is square.

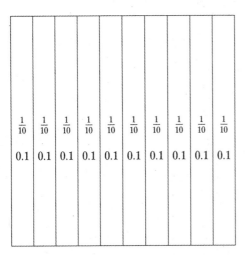

Here are some examples of fractions represented on tenths grids. Fraction names and decimal names for the shaded part are given below each drawing.

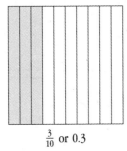

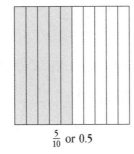

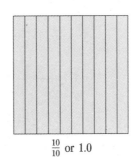

$\frac{3}{10}$ or 0.3 $\frac{5}{10}$ or 0.5 $\frac{10}{10}$ or 1.0

4.2

Making Smaller Parts

At a Glance

Launch

- Review decimal notation with your students.

- Discuss how decimals can be represented on tenths and hundredths grids.

Explore

- As students translate the fractions on their garden plans to decimals, ask questions to guide them into seeing that all the decimal parts must add to 1.

Summarize

- Have a few groups present and explain their solutions.

- As a class, talk about the follow-up questions, and explore thousandths and ten-thousandths grids in greater depth.

- Discuss the base ten number system using the place-value chart. (*optional*)

Assignment Choices

ACE questions 1–9 (9 requires Labsheet 4.ACE) and unassigned choices from earlier problems

We can further divide a tenths grid by drawing horizontal lines to make ten rows. Now we have 100 parts. This *hundredths grid* is what Justin used to plan his garden.

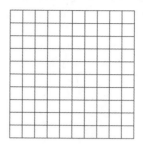

Fractions can also be represented on a hundredths grid. We can write fractional parts of 100 as decimal numbers.

Fraction	Decimal	Meaning	Representation on a hundredths grid
$\frac{7}{100}$	0.07	7 out of 100	
$\frac{27}{100}$	0.27	27 out of 100	
$\frac{51}{100}$	0.51	51 out of 100	

Answer to Problem 4.2

Answers will vary.

Problem 4.2

Look back at the original plan you drew for Justin's garden. Write each of the fractional parts for the vegetables in your plan as a decimal.

Problem 4.2 Follow-Up

1. **a.** What would a hundredths grid look like if each square of the grid were divided into ten equal parts? How many parts would the new grid have?
   b. What is a fraction name for the smallest part of this new grid? A decimal name?
   c. How would you shade an area of this new grid to show $\frac{1}{10}$?
   d. What fraction or decimal names could you call this shaded area?
   e. What would you call this new grid, which has every square of a hundredths grid divided into ten equal parts?

2. **a.** You can write $\frac{9}{100}$ as the decimal 0.09. How could you write $\frac{9}{1000}$ as a decimal?
   b. How could you write $\frac{469}{1000}$ as a decimal?

3. **a.** What would you need to do to the new grid you discovered in question 1 to make a grid that shows *ten thousandths*?
   b. How could you write $\frac{9}{10,000}$ as a decimal?
   c. How could you write $\frac{469}{10,000}$ as a decimal?

4.3 Using Decimal Benchmarks

In Investigation 2, we developed benchmarks to help us estimate fractions. Benchmarks can also help us estimate and compare decimals. You can use what you already know about fractions to make estimating and comparing decimals easier.

Did you know?

Throughout history mathematicians have used many different notations to represent decimal numbers. For example, in 1585, Simon Stevinus would have written 2.57 as either 2, 5' 7" or 2 ⓪ 5 ①7 ②. In 1617, John Napier would have written 2/57. Other commonly-used notations included an underscore, 2<u>57</u>, and a combination of a vertical line and an underscore, 2|<u>57</u>. Even today, the notation varies from country to country. For example, in England, 2.57 is written as 2•57, and, in Germany, it is written as 2,57.

Investigation 4: From Fractions to Decimals 43

At a Glance

Launch

- As a class, develop the set of five decimal benchmarks, and use them to compare pairs of decimals.

Explore

- Circulate, asking questions to help students apply what they learned about fractions to the decimals they are now considering.

Summarize

- Discuss the answers to the problem and the follow-up.
- Extend the discussion, posing more difficult ordering problems for students and focusing on the meaning of decimal place values.

Answers to Problem 4.2 Follow-Up

1. a. 1000 parts
   b. $\frac{1}{1000}$; 0.001
   c. You would still shade $\frac{1}{10}$ of the grid, but now the tenth would be subdivided into 100 parts.
   d. Possible answers: $\frac{100}{1000}$, 0.100, 0.10, 0.1
   e. A thousandths grid

2. a. 0.009
   b. 0.469

3. a. Subdivide each of the thousand sections into 10 parts, which would give 10,000 squares.
   b. 0.0009
   c. 0.0469

Assignment Choices

ACE questions 10–36, 43, 44, and unassigned choices from earlier problems

Playing Distinguishing Digits

At a Glance

Launch

- As a class, investigate the sample Distinguishing Digits puzzle.

- Talk about how the clues relate to place value.

Explore

- Circulate while students solve the puzzles, focusing them on the importance of place value in working with the clues.

Summarize

- As a class, check answers and share puzzle-solving strategies.

- Have groups share the puzzles they created and discuss the strategies they used for developing their puzzles. (optional)

Assignment Choices

Unassigned choices from earlier problems

Problem 4.3

A. Rename each of these fraction benchmarks as a decimal.

  1. 0   **2.** $\frac{1}{4}$ **3.** $\frac{1}{2}$ **4.** $\frac{3}{4}$ **5.** 1

B. Now use the decimal benchmarks and other strategies that make sense to you to help you order each set of numbers from smallest to largest.

1. 0.23	0.28	0.25
2. 2.054	20.54	2.54
3. 0.78	0.708	0.078

C. For each of the three decimals in parts 1 and 3 of question B, give the name of the decimal in words, and tell which benchmark the number is nearest. Give the benchmark as a fraction and as a decimal. For each decimal benchmark chosen, explain your reasoning. Organize your work in a table like the one below.

Number	Name in words	Nearest decimal benchmark	Nearest fraction benchmark	Reasoning
0.23				
0.28				
0.25				
0.78				
0.708				
0.078				

Problem 4.3 Follow-Up

1. Write, as a decimal, a fraction with a denominator of 10,000 that is near, but less than, $\frac{3}{4}$.

2. Write, as a decimal, a fraction with a denominator of 1000 that is near, but less than, 1.

4.4 Playing Distinguishing Digits

Distinguishing Digits is a collection of number puzzles. In each puzzle, you use clues to help find a Mystery Number.

Answers to Problem 4.3

A. **1.** 0   **2.** 0.25   **3.** 0.5   **4.** 0.75   **5.** 1

B. **1.** 0.23   0.25   0.28
    2. 2.054   2.54   20.54
    3. 0.078   0.708   0.78

C. See page 52h.

Answers to Problem 4.3 Follow-Up

1. Answers will vary. Possible answers: $\frac{7398}{10,000}$, $\frac{7450}{10,000}$, $\frac{7499}{10,000}$

2. Answers will vary. Possible answers: $\frac{999}{1000}$, $\frac{987}{1000}$, $\frac{970}{1000}$

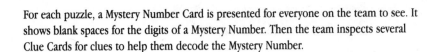

For each puzzle, a Mystery Number Card is presented for everyone on the team to see. It shows blank spaces for the digits of a Mystery Number. Then the team inspects several Clue Cards for clues to help them decode the Mystery Number.

Look at the following example. As a class, decide what the Mystery Number must be.

Mystery Number

1 __ . __ __ __

Clue 1 The digit in the thousandths place is double the digit in the ones place.
Clue 2 The digit in the tenths place is odd, and it represents the sum of the digits in the tens place and the thousandths place.
Clue 3 There are exactly two odd digits in the Mystery Number.
Clue 4 The digit in the hundredths place is three times the digit in the ones place.

Problem 4.4

Play the Distinguishing Digits puzzles with your group. Record the strategies you use to solve the puzzles.

■ Problem 4.4 Follow-Up

With your group, create a new Distinguishing Digits puzzle. Try out the puzzle on another group. If the other group finds a problem with your puzzle, rework the clues until your puzzle works.

Answers to Problem 4.4

Note that some of the puzzles have more than one answer.

Puzzle 1: 842
Puzzle 2: 513
Puzzle 3: 826,413 and 862, 431
Puzzle 4: 0.248
Puzzle 5: 0.132 and 0.396
Puzzle 6: 0.468 and 0.648
Puzzle 7: 0.3609
Puzzle 8: 216.12
Puzzle 9: 5612.034
Puzzle 10: 44,527.42986

Answers

Applications

1. $\frac{30}{100}$, $\frac{3}{10}$, 0.30, 0.3
2. $\frac{32}{100}$, $\frac{16}{50}$, $\frac{8}{25}$, 0.32
3. $\frac{53}{100}$, 0.53
4. $\frac{60}{100}$, $\frac{30}{50}$, $\frac{15}{25}$, $\frac{3}{5}$, 0.60, 0.6
5. $\frac{90}{100}$, $\frac{45}{50}$, $\frac{9}{10}$, 0.90, 0.9
6. $\frac{75}{100}$, $\frac{15}{20}$, $\frac{3}{4}$, 0.75

As you work on these ACE questions, use your calculator whenever you need it.

Applications

In 1–6, the hundredth grid is partially shaded. Write fraction and decimal names to describe the shaded part.

1.
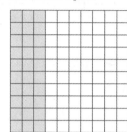

2.
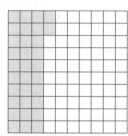

3.
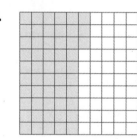

4.
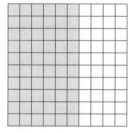

5.
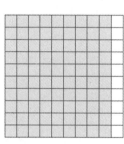

6.
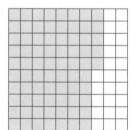

In 7–8, the whole is one hundredth grid. Write fraction and decimal names to describe the shaded part.

7.

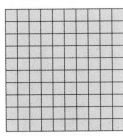 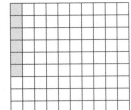

8.

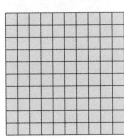 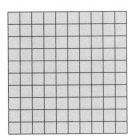 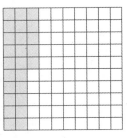

9. In a–f, use the blank hundredths grids on Labsheet 4.ACE to shade the given fractional part. Write the fraction as an equivalent decimal.

a. $\frac{1}{2}$ of the hundredths grid

b. $\frac{3}{4}$ of the hundredths grid

c. $\frac{99}{100}$ of the hundredths grid

d. $1\frac{3}{10}$ of the hundredths grids

e. $2\frac{7}{10}$ of the hundredths grids

f. $1\frac{3}{5}$ of the hundredths grids

In 10–13, rewrite each pair of numbers, inserting a less-than symbol (<), a greater-than symbol (>), or an equals symbol (=) between the numbers to make a true statement.

10. $\frac{3}{5}$ 0.3

11. 0.205   0.21

12. 0.1   0.1000

13. $\frac{37}{50}$ 0.74

Investigation 4: From Fractions to Decimals   **47**

7. $1\frac{6}{100}$, $1\frac{3}{50}$, 1.06

8. $2\frac{25}{100}$, $2\frac{5}{20}$, $2\frac{1}{4}$, 2.25

9a. See below left.

9b. See below left.

9c. See page 52i.

9d. See page 52i.

9e. See page 52i.

9f. See page 52j.

10. $\frac{3}{5}$ > 0.3

11. 0.205 < 0.21

12. 0.1 = 0.1000

13. $\frac{37}{50}$ = 0.74

9a. 0.50 or 0.5; Possible answer:

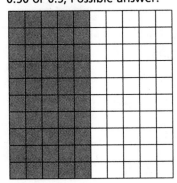

9b. 0.75; Possible answer:

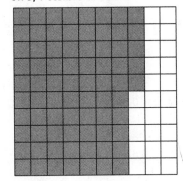

14. 0.12, 0.127, 0.2, 0.33,
$\frac{45}{10}$

15. $\frac{3}{1000}$, 0.005, $\frac{3}{100}$, 0.34

16. 0.827, $\frac{987}{1000}$, 1.23, $\frac{987}{100}$

17.–20. See below right.

21. 0.3 < 0.6

22. 0.4 < $\frac{3}{5}$

23. 0.7 > $\frac{1}{2}$

In 14–16, rewrite the numbers in order from smallest to largest.

14.   0.33   0.12   0.127   0.2 $\frac{45}{10}$

15. $\frac{3}{100}$ 0.005 $\frac{3}{1000}$ 0.34

16.   0.827   1.23 $\frac{987}{100}$ $\frac{987}{1000}$

In 17–20, copy the part of the number line given. Then, find the "step" by determining the difference from one mark to another. Label the unlabeled marks with decimal numbers. The first step is given for you.

17. ◄——┼————┼————┼————┼——►   The step is 0.2.
       0.2   0.8

18. ◄——┼————┼————┼————┼——►   The step is _____.
       0.15   0.17

19. ◄——┼————┼————┼————┼——►   The step is _____.
       0.028   0.029

20. ◄——┼————┼————┼————┼——►   The step is _____.
       1.8   1.9

In 21–35, rewrite each pair of numbers, inserting a less-than symbol (<), a greater-than symbol (>), or an equals symbol (=) between the numbers to make a true statement.

21.   0.3   0.6

22.   0.4 $\frac{3}{5}$

23.   0.7 $\frac{1}{2}$

17.   The step is 0.2.

18.   The step is 0.01.

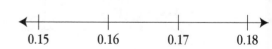

19.   The step is 0.001.

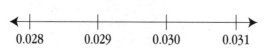

20.   The step is 0.1.

24. 0.34   0.23

25. 0.60   0.6

26. 0.52 $\frac{2}{4}$

27. 0.34   0.4

28. 0.08   0.8

29. 0.92   0.9

30. 2.45   2.3

31. 0.56   0.056

32. 0.037   0.029

33. 0.7   0.725

34. 0.41   0.405

35. 0.10   0.108

Connections

36. Su computed her free-throw average on her calculator and got 0.6019. Ahmed computed his free-throw average and got 0.602. Solange's free-throw average was 0.62. Who is the best free-throw shooter? Explain your answer.

37. Chad, Roman, and Kari wanted to know who was the best free-throw shooter among them. Chad's free-throw average was about 0.588. Roman's average was close to 0.611. Kari consistently made 6 out of 10. Who is the best free-throw shooter? Explain your answer.

In 38–40, show the fraction by drawing and shading squares. Then, write each improper fraction as an equivalent mixed number.

38. $\frac{7}{5}$ **39.** $\frac{7}{3}$ **40.** $\frac{11}{6}$

38. $\frac{7}{5} = 1\frac{2}{5}$. Possible illustration:

 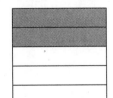

---

24. 0.34 > 0.23

25. 0.60 = 0.6

26. 0.52 > $\frac{2}{4}$

27. 0.34 < 0.4

28. 0.08 < 0.8

29. 0.92 > 0.9

30. 2.45 > 2.3

31. 0.56 > 0.056

32. 0.037 > 0.029

33. 0.7 < 0.725

34. 0.41 > 0.405

35. 0.10 < 0.108

Connections

36. Solange is the best free-throw shooter. Possible explanation: 0.62 is the same as 0.6200, and 0.602 is the same as 0.6020. Now that all three decimals have the same number of decimal places, you can just compare the digits. Since 6200 > 6020 > 6019, 0.62 is the largest decimal.

37. Roman is the best free-throw shooter. Possible explanation: Kari's average can be written as 0.600. Since 611 > 600 > 588, 0.611 is the largest decimal.

38. See below left.

39. See page 52j.

40. See page 52j.

41a. $\frac{9}{16}$

41b. $\frac{8}{25}$

41c. $\frac{9}{16}$

41. Determine what part of each figure is shaded.

a.

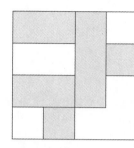

b.

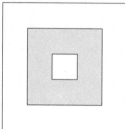

c.

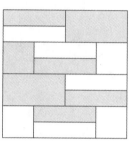

Extensions

Problems 42–43 involve the Dewey Decimal system, which is used in many libraries to catalog books. The Dewey Decimal system is based on the decimal number system.

42. Serita looked up *elephant* in the library's computer. She saw that there were several books about elephants, with these call numbers: 599.55, 599.504, 599.5, and 599.5044. The librarian showed Serita the guide numbers on the ends of the shelves and explained that the books were arranged in numerical order from smallest to largest and from left to right. In what order will the books about elephants be arranged on the shelf? Explain your reasoning.

43. Huang wanted to reshelve two books he had been reading about the history of rock and roll music. The books' call numbers were 782.42 and 781.66. Huang located the shelf where his book belonged, but the other books he found there were completely out of order: 781.5, 782.005, 781.053, 781.035, and 782.409. He rearranged the books already on the shelf and then placed his books among them. In what order did Huang put the books that were already on the shelf? Where did he put his books?

44. Microchips can process information in as little as one billionth of a second, which is called a *nanosecond*. Write the decimal representation of a nanosecond.

Extensions

42. 599.5, 599.504, 599.5044, 599.55. Possible explanation: You can think of each call number as having four digits after the decimal place: 599.5500, 599.5040, 599.5000, 599.5044. Now the decimal parts can be ordered as if they were whole numbers: 5000, 5040, 5044, and 5500.

43. The books already on the shelf should be arranged, from left to right, in this order: 781.035, 781.053, 781.5, 782.005, and 782.409. Huang's book with call number 781.66 belongs in between 781.5 and 782.005. His book with call number 782.42 belongs at the end of all the books.

44. A nanosecond can be written as 0.000000001 seconds.

Possible Answers

1. You could divide it into two strips in one direction and five strips in the other. Or, you could divide it into ten vertical or horizontal strips.

To represent hundredths, you could divide each section of the square representing tenths into ten equal parts.

To represent thousandths, you could divide each section of the square representing hundredths into ten equal parts. That would give you 1000 equal parts.

2. First, see whether the whole numbers to the left of the decimal point are different. If they are, then the larger whole number will determine which decimal number is larger. If they are the same, rewrite each decimal so there are the same number of digits after the decimal. Then, the digits can be compared, starting with the tenths place, until one is found to have a larger place value.

3. Reading from left to right, the 2 represents two ten thousands, the 5 represents five thousands, the 7 represents seven hundreds, the 0 represents zero tens, the 8 represents eight ones, the 2 represents two tenths, the 0 represents zero hundredths, and the 1 represents one thousandth.

4. See page 52j.

Mathematical Reflections

In this investigation, you used square grids to help you model decimals. You found that decimals are special kinds of fractions. These questions will help you summarize what you have learned:

(1) Describe the process of dividing a square to represent tenths; to represent hundredths; to represent thousandths.

(2) When comparing two decimals, how can you decide which decimal represents a larger number?

(3) Decimals are an extension of the place-value system you studied for whole numbers in elementary school. Tell what each digit in 25,708.201 represents.

(4) Express each of the following decimal numbers in words.

0.16   1.069   33.109   431.0115

Think about your answers to these questions, discuss your ideas with other students and your teacher, and then write a summary of your findings in your journal.

Tips for the Linguistically Diverse Classroom

Diagram Code The Diagram Code technique is described in detail in *Getting to Know CMP.* Students use a minimal number of words and drawings, diagrams, or symbols to respond to questions that require writing. Example: Question 2—A student might answer this question by writing 2.21 < 3.99, Why? 2.21, 3.99, 2 < 3; 3.3 > 3.20, Why? 3.30, 3. 20, 30 > 20.

4.1 • Designing a Garden

Launch

This is a good problem for students to work on individually or in pairs. Each student or pair of students will need one or more copies of Labsheet 4.1. Each square on the grid represents 1 square meter of the 100-square-meter plot.

Introduce the problem by telling the story of the family garden plot.

> You must determine how many square meters of the garden area need to be planted with each vegetable. Justin's father wants twice as much of the garden to be planted with corn as potatoes (that is, for every 1 square meter planted with potatoes, there will be 2 square meters planted with corn). He also wants three times as much land planted with potatoes as tomatoes (that is, for every 1 square meter planted with tomatoes, there will be 3 square meters planted with potatoes). This means that the numbers of square meters planted in corn, potatoes, and tomatoes are linked.

Students will have to determine relationships that work; for example, for 1 square meter of tomatoes, there will be 3 square meters of potatoes and 6 square meters of corn.

> The remaining square meters can be used for the other vegetables that the family wants planted.

Be sure students understand that their designs must satisfy all of the requirements of Justin's family. Beginning with the grid may be too abstract for many students. Consequently, you may want to give each team of students 100 colored tiles or cubes, which will help them distinguish different sections of the plot they are to make. They can translate their finished plan to the grid.

Explore

Circulate and ask questions, refocusing students who are off-task. Encourage students who are struggling. Each group should be able to show how their garden design satisfies the conditions.

You may want to have a few groups translate their plan to copies of Transparency 4.2D to make sharing with the class easier. If some groups finish early, ask them to work on the follow-up questions.

Summarize

Let a few groups show their garden plans. Ask each group for proof that their design fits the necessary conditions.

Raise the follow-up questions of the largest possible amount of land planted in potatoes and the smallest possible amount planted in potatoes. Have groups share the largest and the smallest they found, and challenge the class to see whether they can make a design that fits the conditions with fewer or more square meters in potatoes than any found so far.

Launch

We assume students have met decimals before, but you will want to review them. If your students have trouble reading decimals, remind them that decimals are just easy ways to write fractions with denominators of 10, 100, 1000, 10,000, and other powers of ten. The chart in the "For the Teacher" box on page 52e will be helpful.

In this problem, students are introduced to tenths and hundredths using grid models. The grids on Transparencies 4.2C and 4.2D can make the class discussion of tenths and hundredths easier. In moving from the tenths grid to the hundredths grid, encourage students to focus on the fact that they are moving from a grid of 10 columns to one on which horizontal lines have been added to make 10 rows, resulting in 100 squares. Highlight how shading on the tenths grid and hundredths grid compare. In particular, note that 0.5 and 0.50 represent the same area of the grid.

> Can you show 0.3 and 0.4 on a hundredths grid? Why or why not? (*Yes, because tenths can be rewritten as hundredths.*) Can you show 0.63 and 0.36 on a tenths grid? Why or why not? (*No, because while 0.30 and 0.40 can be rewritten as tenths, 0.63 and 0.36 cannot. We need smaller divisions of the grid to show these amounts.*)

Now, read the problem with the students.

> The problem asks what decimal amount you planned for each vegetable in Justin's garden. You need to represent the fractional amount for each area as a decimal.

Let students work on Problem 4.2 individually or in their groups from Problem 4.1.

Explore

As you circulate, ask what the fractions for each type of vegetable should add to. Many students will need several opportunities to see that the sum of all the fractional parts must equal the whole, which is 1. You may want a couple of groups to add their decimal amounts to the copy of Transparency 4.2D that they prepared for Problem 4.1 for sharing with the class.

Summarize

Have a few groups share their answers with the class. Discuss the follow-up questions, which encourage students to visualize further dividing the hundredths grid to create thousandths and ten-thousandths grids, and to name parts of these grids with decimals and fractions.

Give each student a copy of Labsheet 4.2, which contains a large hundredths grid. Have students do some initial brainstorming and sketching in their groups about how to create a thousandths grid from the hundredths grid. Then, as a class, discuss what it would mean to show each square divided into ten parts.

Each square on the grid would be a replica of the tenths grid, only much smaller. You may want to demonstrate dividing one of the hundredth squares on Transparency 4.2D into ten equal parts.

> What would happen if each of the hundred squares were divided in this way? How many divisions would there be? (*1000*) What would we call this grid? (*a thousandths grid*)

In one classroom, the following discussion occurred.

Teacher In your book on page 41, you see a square that is divided into ten parts to show tenths. Could someone tell me how I could take this square and divide it into tenths like what is shown in the book? (*The teacher displayed Transparency 4.2B.*)

Ted You could divide the square in half by drawing a line down the middle, and then divide each half into five equal parts by drawing more lines from the top to the bottom.

Teacher How do I know where to draw the lines to divide each half into five equal parts?

Ted You could measure each half and divide that by 5 to see how wide to make each part.

Teacher Is that reasonable? Do others think that will give me a square divided into ten equal parts? (*The teacher and class determined where to place the lines.*)

In your book on page 42 is a square that has been divided into hundredths. Could someone tell me how I could take our square, which is now divided into tenths, and further divide it to make a square showing hundredths?

Latisia Turn the square so that the lines are going across, and then do the same thing Ted told you to do before.

Teacher Is that reasonable? Do others think that will give me a square divided into 100 equal pieces? (*The teacher followed Latisia's directions.*)

Suppose you wanted to show each of these hundredth squares divided into ten equal parts. How many parts would the new grid have? How would you divide each of the squares into ten equal parts?

I will give each of you a hundredths grid that looks similar to what I have on the overhead. I want you to divide a couple of the hundredths squares into ten equal parts and then share what you have done with the person next to you. See if you both agree that you have done what was asked, and then discuss how many parts the grid would have if you continued this process. (*The teacher noticed that several students were struggling to divided a hundredths square into ten parts. Many divided squares into fourths and then didn't know how to proceed.*) Think about the strategy Ted used to divide the large original square. Can you use that strategy here?

Having students begin to divide the hundredths grid to get thousandths was worthwhile, because several misconceptions became clear and the teacher was able to address them in the class discussion. Also, students got a better feel for how thousandths relate to hundredths and to tenths.

Ask questions to assess what your students are understanding about representing quantities as decimals, and have students discuss how to write numbers involving thousandths as decimals.

Can you show 0.3 on a tenths grid? On a hundredths grid?
A thousandths grid? Why or why not?

Can you show 0.34 on a tenths grid? On a hundredths grid? A
thousandths grid? Why or why not?

Can you show 0.345 on a tenths grid? On a hundredths grid?
A thousandths grid? Why or why not?

In follow-up question 3, students think about a ten-thousandths grid. You can use Transparencies
4.2E and 4.2F to help students better visualize moving from the thousandths to the ten-thousandths grid.

What would you need to do to the thousandths grid to make a grid
with 10,000 squares? Can you see that this is like creating 100 squares in
each of the squares on a hundredths grid?

When students understand the ten-thousandths grid, you may want to extend the ideas in the
follow-up. Shade in decimal amounts on the grids as you consider various fractions, using copies
of Transparencies 4.2C, 4.2D, 4.2E, and 4.2F.

Can we show 0.3 and 0.8 on a ten-thousandths grid? Can we show 0.34
and 0.89 on a ten-thousandths grid? Can we show 0.345 and 0.897 on a
ten-thousandths grid? Can we show 0.3452 and 0.8974 on a ten-thousandths grid? On a thousandths grid? A hundredths grid? A tenths grid?
Why or why not?

For the Teacher

You may want to make a wall display or a bulletin board of the place-
value chart on the next page and work with students to think about
what it means to extend the chart in either direction. This chart is also
shown on Transparency 4.2G.

The system we use for writing numbers is called the *base ten number system.* It uses groups of ones, tens, hundreds, thousands, ten thousands,
and so on. For example, 69 represents 6 groups of ten and 9 groups of
one; 28,590 represents 2 groups of ten thousand, 8 groups of one thousand, 5 groups of one hundred, 9 groups of ten, and 0 groups of one.

Over time, people realized they needed to extend the number system to
represent numbers smaller than 1. A decimal point separates these digits
so that the numbers to the right of the decimal point represent fractions
whose denominators are ten (tenths), one hundred (hundredths), one
thousand (thousandths), ten thousand (ten thousandths), and so on. For
example, 5.8 represents 5 groups of one and 8 groups of one tenth;
36.420 represents 3 groups of ten, 6 groups of one, 4 groups of one
tenth, 2 groups of one hundredth, and 0 groups of one thousandth.

(continued)

Ask questions to help students interpret the chart.

> Do you see the symmetry of the names of the place values about the 1? What are your predictions for the next few lines of the chart if you were to extend it both to the left and to the right?

> Choose a number from the chart. What relationships do you see between it and the numbers directly to its right and left?

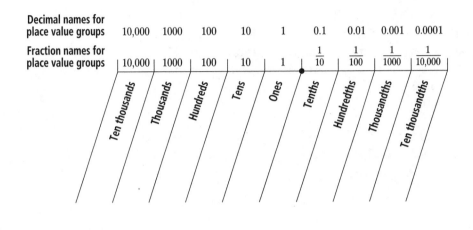

4.3 • Using Decimal Benchmarks

Launch

This problem gives students an opportunity to connect the fraction benchmarks $0, \frac{1}{4}, \frac{1}{2}, \frac{3}{4},$ and 1 to decimal benchmarks. Launch the problem by exploring an example.

> For decimals to be useful to us, we need to be able to estimate how large a number is and to make comparisons between numbers. Remember that we did this for fractions, so we can build on what we know about fractions. Here are two decimal numbers: 0.05 and 0.3. (*Write these on the board.*) Which of these represents the greater quantity? How do you know?

Some students have a hard time giving up on their whole-numbers idea that 5 is larger than 3. Having someone shade grids at the overhead, perhaps using Transparency 4.2D, to show the two values can be helpful, and gives you a chance to reinforce the meaning of decimals and how to read them.

> Who can give me a fraction name for each of these numbers? $(\frac{5}{100}, \frac{3}{10})$ Tell me how you found the fraction equivalents.

When we looked at benchmarks with fractions, we used 0, $\frac{1}{2}$, and 1. Using these benchmarks, which number is $\frac{5}{100}$ closest to? (*0*) Which number is $\frac{3}{10}$ closest to? ($\frac{1}{2}$) Explain your answers.

Let's add more benchmarks to our set: $\frac{1}{4}$ and $\frac{3}{4}$.

The problem asks students to find decimal representations for these new benchmarks. You may want to help your students to convert the new fraction benchmark set of 0, $\frac{1}{4}$, $\frac{1}{2}$, $\frac{3}{4}$, and 1 into decimal benchmarks: 0, 0.25, 0.5, 0.75, and 1. Then write the fraction benchmarks 0, $\frac{1}{4}$, $\frac{1}{2}$, $\frac{3}{4}$, and 1 on the board with their decimal equivalents underneath.

With this new benchmark set, is $\frac{5}{100}$ closer to 0 or $\frac{1}{4}$? (*0*) Is $\frac{3}{10}$ closer to $\frac{1}{4}$ or $\frac{1}{2}$? ($\frac{1}{4}$) How do you know? What is the decimal representation for $\frac{1}{4}$? (*0.25*) How do you know? If we say that $\frac{5}{100}$ is closer to 0 than $\frac{1}{4}$, then 0.05 must be closer to 0 than 0.25. If $\frac{3}{10}$ is closer to $\frac{1}{4}$ than $\frac{1}{2}$, then 0.3 must be closer to 0.25 than 0.5.

If your students are struggling, you might do another example together. You could choose decimals on either side of $\frac{1}{2}$, such as 0.52 and 0.499. This pair again raises the issue of whether we are comparing 52 and 499 or the fractions $\frac{52}{100}$ and $\frac{499}{1000}$.

When your students seem to understand the task, let them work in pairs or groups on Problem 4.3.

Explore

As you visit the groups, ask questions that encourage students to use the reasoning they developed for fractions to help estimate the size of the numbers they are now investigating. As students look for ways to compare decimals directly, don't let the process proceed to an algorithm too soon. Students will be better served if the focus stays on, Why is this true? rather than on, How do I do this?

Summarize

A quick check of answers and a discussion of strategies will make a good beginning to the summary. Then you can pose a harder ordering problem for students using a think-pair-share strategy. Have students think individually about the problem you pose, then share their ideas in pairs and try to reach consensus. For example, have students order these numbers from smallest to largest:

$$0.827 \qquad \frac{827}{100} \qquad 1.827 \qquad \frac{1827}{10,000} \qquad 0.087$$

One way to tackle the problem is to rewrite the numbers as decimals:

$$0.827 \qquad 8.27 \qquad 1.827 \qquad 0.1827 \qquad 0.087$$

When pairs have considered the problem, discuss it as a class.

Which two of these numbers are greater than a whole? (*8.27 and 1.827*) Which of these is larger, and how do you know? (*Here the whole number part tells the story: 8.27 is larger.*) The three remaining numbers are 0.827, 0.1827, and 0.087. Which of these is the largest?

You want the conversation to center on the meaning of the *places* in the numbers.

> Where is the tenths place? (*to the right of the decimal point*) The hundredths place? (*two places to the right of the decimal point*) Which place has the larger value? (*the tenths place*) Which of the three remaining numbers has the greatest number of tenths? (*0.827, with 8 tenths, then 0.1827, then 0.087*) This gives us our ordering:
>
> 0.087   0.1827   0.827   1.827   8.27
>
> Who can tell me a decimal number between each of these pairs of numbers?

From this summary, your students should have a good set of strategies for making sense of and comparing decimals.

4.4 • Playing Distinguishing Digits

Launch

The best way to launch the Distinguishing Digits puzzle is to investigate the sample puzzle given in the student edition. From the beginning, stress that proving the solution is part of successfully solving the puzzle.

> Look at the Distinguishing Digits puzzle in your books. You will see a Mystery Number, with blanks for the missing digits, and a set of clues about the number. Your challenge is to use the clues to figure out what numerals go in the blanks to make a number that satisfies all of the clues.
>
> First, look at the skeleton for the Mystery Number. What can you tell me about the number? (*It is between 10 and 20. It has three decimal places, so it is a whole number plus some number of thousandths.*)
>
> Look at the first clue. What can you tell me from this information? (*The digit in the thousandths place is 2, 4, 6, or 8 and the digit in the ones place is 1, 2, 3, or 4*)

Be sure students specifically locate the place values to which the clue relates.

> Look at all the clues, and tell me what you know. It is not necessary to use the clues in the order they are given. In fact, it is often helpful to start somewhere else. Look over the clues to decide which convey the simplest information.

Students will have different beginning ideas. For instance, Clue 3 indicates that we can use only one other odd digit, since one odd digit is already given (the 1 in the tens place). The first part of the second clue tells us where the second odd digit will go.

Next, knowing that all of the remaining digits must be even, the last clue is easier to decipher. Make a list of pairs of even numbers in which the second number is three times the first; the only set of single-digit numbers that works is 2 and 6. This gives us part of our answer:

12. odd 6_____

Now consider the unused clues. Since the ones place is filled, we can employ Clue 1: the digit in the thousandths place is 4, which leads us to complete the last part of Clue 2. The sum of the digits in the tens place and the thousandths place is 5, giving the solution of 12.564.

Divide students into groups. In solving puzzles, students are sometimes more fully engaged if all students in the group are near the same level of understanding. However, students should have an opportunity to hear different strategies and ways of reasoning, so you may want to begin with a mixture of levels of understanding and have a second round pairing students of more similar understanding. Pass out the envelopes of Mystery Numbers cards and Clue cards to the groups. You may want to have every group try every puzzle, or give every group a few puzzles to solve.

Explore

As you circulate, ask questions about place value. Monitor whether every student is having an opportunity to contribute. When students have finished solving the puzzles, have them work on the follow-up.

Summarize

Give the class a chance to check answers and to share strategies.

You might have students share some of the problems they developed for the follow-up. Ask what they had to think about to develop a set of clues that would work. Developing the puzzles and talking about their strategies will enhance students' understanding of place value.

Additional Answers

Answers to Problem 4.3

C.

Number	Name in words	Nearest decimal benchmark	Nearest fraction benchmark	Reasoning
0.23	twenty-three hundredths	0.25	$\frac{1}{4}$	$\frac{23}{100}$ is very close to $\frac{25}{100}$ or $\frac{1}{4}$.
0.25	twenty-five hundredths	0.25	$\frac{1}{4}$	$\frac{25}{100}$ equals $\frac{1}{4}$.
0.28	twenty-eight hundredths	0.25	$\frac{1}{4}$	$\frac{28}{100}$ is close to $\frac{25}{100}$, or $\frac{1}{4}$.
0.78	seventy-eight hundredths	0.75	$\frac{3}{4}$	$\frac{78}{100}$ is close to $\frac{75}{100}$, or $\frac{3}{4}$.
0.708	seven hundred and eight thousandths	0.75	$\frac{3}{4}$	$\frac{708}{1000}$ is close to $\frac{700}{1000}$ or $\frac{70}{100}$. $\frac{70}{100}$ is close to $\frac{75}{100}$.
0.078	seventy-eight thousandths	0	0	$\frac{78}{1000}$ is close to $\frac{80}{1000}$, or $\frac{8}{100}$. $\frac{8}{100}$ is close to zero.

ACE Questions

Applications

9c. 0.99; Possible answer:

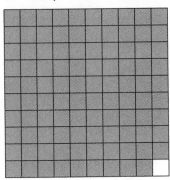

9d. 1.30 or 1.3; Possible answer:

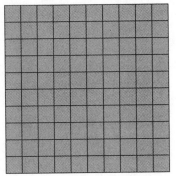 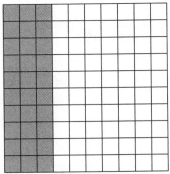

9e. 2.70 or 2.7; Possible answer:

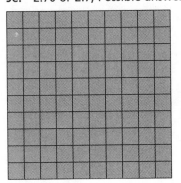 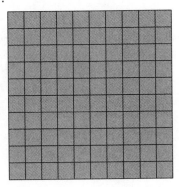 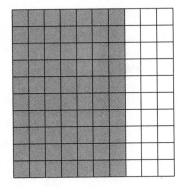

9f. 1.60 or 1.6; Possible answer:

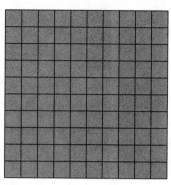

Connections

39. $\frac{7}{3} = 2\frac{1}{3}$. Possible illustration:

  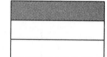

40. $\frac{11}{6} = 1\frac{5}{6}$. Possible illustration:

 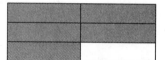

Mathematical Reflections

4. 0.16 is sixteen hundreths; 1.069 is one and sixty-nine thousandths; 33.109 is thirty-three and one hundred nine thousandths; 431.0115 is four hundred thirty-one and one hundred fifteen ten thousandths.

Moving Between Fractions and Decimals

Mathematical and Problem-Solving Goals

- **To understand that the decimal representation of a fraction shows the same proportion but is based on a power of 10 as the denominator**

- **To use the concept of equivalence to change the form of simple fractions to fractions with 100 in the denominator**

- **To understand the division interpretation of fractions and use it to change fractions to decimals**

- **To use fraction strips to estimate fractions as decimals and decimals as fractions**

- **To find equivalent forms of fraction quantities**

- **To use hundredths grids to model fractions**

- **To use knowledge of operations, fractions, and decimals to understand real-world situations**

Fractions are especially difficult for students because there are several ways to interpret them. In Investigations 1 and 2, students focused on the part-whole interpretation using fraction strips and number lines as models. In Investigation 3, this model was expanded to include areas of figures, especially rectangles. In Investigation 4, this area model was applied to help students make sense of decimal fractions. In this investigation we develop the meaning of fractions as implied division, an interpretation that allows students to use a calculator to change fractions to decimals, facilitating comparisons and computations.

Problem 5.1, Choosing the Best, focuses on making comparisons among three quantities that can be represented with fractions. In Problem 5.2, Writing Fractions as Decimals, students use fraction strips, including a hundredths strip, to estimate fraction and decimal equivalents. The goal is to help students focus on fractions and decimals as quantities that can be represented in more than one form. In Problem 5.3, Moving Between Fractions and Decimals, whole number division helps extend the meaning of fractions as implied divisions. This problem helps students to understand why a fraction can be interpreted as an implied division and to use implied division to change fractions to decimal representations.

Materials

For students

- Labsheets 5.1 and 5.2 (1 per student)
- Colored tiles or other manipulatives (optional)
- Scissors
- Straightedges
- Chart paper, or a transparency of Labsheet 5.2 and transparency markers (optional; for recording answers to share with the class)

For the teacher

- Transparencies 4.2D, 5.1, 5.2, and 5.3 (optional)
- Transparency of Labsheet 5.2 (optional)

INVESTIGATION 5

Moving Between Fractions and Decimals

In your daily life, you often need to choose among options. You might have to decide which option is the best buy, gives the best outcome, or yields the most money. In these situations, mathematics can help you make comparisons so you can make a good decision.

5.1 ## Choosing the Best

The Portland Middle School basketball team is playing the Coldwater Colts. The game is tied 58 to 58. In her excitement the Coldwater coach steps onto the court just as the buzzer sounds, and a technical foul is called.

The Portland coach has to choose one of her players to shoot the free throw. If the player makes the free throw, Portland will win.

Did you know?

Basketball was invented in 1891 by James Naismith, a physical education teacher who wanted to create a team sport that could be played indoors during the winter. The game was originally played with a soccer ball, and peach baskets were used as goals.

5.1

Choosing the Best

At a Glance

Launch

- Tell the story of the coach's choice.
- Make sure students understand that they need to compare the players' scores in a way that accounts for the varying numbers of trials.

Explore

- If some groups are having difficulty, suggest visual models they may use.
- Pose extension questions for groups that finish early.

Summarize

- Allow groups to present their arguments.
- Make sure both the fraction-strip method and the equivalent-fraction method for solving the problem are discussed.

Assignment Choices

ACE questions 17–24, 39, and unassigned choices from earlier problems

5.2

Writing Fractions as Decimals

At a Glance

Launch

- Present the problem, and work through an example or two with the class.

- Make sure both ways of working with Labsheet 5.2—using a straightedge and maneuvering the hundredths strip—are discussed.

Explore

- As you observe, have students find equivalents in the opposite direction as well: converting decimals to fractions.

Summarize

- Have groups present their findings and explain their reasoning.

- As a class, analyze the strategies and the answers.

- Ask questions to focus students on the decimal-to-fraction change.

Assignment Choices

ACE questions 1–16, 35–38, and unassigned choices from earlier problems

---

Problem 5.1

The coach has three players to choose from to shoot the free throw. In their pregame warm-ups:
- Angela made 17 out of 25 free throws
- Emily made 15 out of 20 free throws
- Carma made 7 out of 10 free throws

Which player should the Portland coach select to shoot the free throw? Explain your reasoning.

Problem 5.1 Follow-Up

Of the top four free-throw shooters on the Coldwater Colts:
- Naomi averages 19 out of 25 free throws
- Bobbie averages 8 out of 10 free throws
- Kate averages 36 out of 50 free throws
- Olympia averages 16 out of 20 free throws

If you were the coach of the Colts, which player would you choose to take the free-throw on a technical foul? Explain your reasoning.

5.2 Writing Fractions as Decimals

The Portland coach chose Emily to take the shot. Emily missed the free throw, but the team won in overtime.

The next day, the players asked their math teacher, Mr. Martinez, what he thought about the problem of whom to choose to take the free throw. He said he always tries to find a decimal name for fractions whose values he needs to compare.

Mr. Martinez explained, "Our fraction strips can help us find decimal names that are good approximations for some fractions. Decimals are ways to express fractions with denominators of 10 or 100. We can use our tenths strip to help us find decimal approximations. We could find even closer approximations if we had a fraction strip divided into hundredths."

Answer to Problem 5.1

Angela's success rate is $\frac{68}{100}$, Emily's is $\frac{75}{100}$, and Carma's is $\frac{70}{100}$. Emily is the most successful shooter and should be chosen to attempt the free throw.

Answer to Problem 5.1 Follow-Up

Naomi averages 76 out of 100 free throws, Bobbie averages about 80 out of 100, Kate averages 72 out of 100, and Olympia averages 80 out of 100 free throws. Most students will probably choose either Bobbie or Olympia to shoot the free throw. They both have the same success rate.

Problem 5.2

On the next page are the fraction strips with which you are already familiar. Below these strips is a hundredths strip, which is a tenths strip that has each segment divided into ten parts. Work with your group to find a way to use the fraction strips to help you estimate each of the fractions represented on the halves, thirds, fourths, fifths, sixths, eighths, ninths, tenths, and twelfths fraction strips as decimals.

You might think about doing this by comparing the marks on each fraction strip to marks on the hundredths strip. For example, to find a decimal name for $\frac{5}{12}$, you can find the mark on the hundredths strip that is nearest to the length $\frac{5}{12}$, since hundredths can easily be written as decimals. Since the mark at $\frac{42}{100}$ on the hundredths strip is the closest mark to $\frac{5}{12}$ on the twelfths strip, $\frac{5}{12}$ is approximately equal to 0.42. Sometimes it is easier to look at the tenths strip. For example, $\frac{1}{2}$ on the fraction strip is at the same mark as $\frac{5}{10}$ on the tenths strip, so $\frac{1}{2}$ is equivalent to 0.5.

On Labsheet 5.2, label each mark on the halves, thirds, fourths, fifths, sixths, eighths, ninths, tenths, and twelfths fraction strips with an approximate decimal representation. Be prepared to explain your answers.

■ Problem 5.2 Follow-Up

1. Did you find any patterns that helped you to predict what some of the fractions would be as decimals?

2. How did your knowledge of equivalent fractions help you to find decimal names for some of your fractions?

3. In a–d, find an approximate fraction for the decimal.
   a. 0.17
   b. 0.29
   c. 0.609
   d. 0.92

Answer to Problem 5.2

See page 66h.

Answers to Problem 5.2 Follow-Up

1. Possible answer: The halves, fifths, and tenths strips are easy because the marks on these strips match exactly with marks on the hundredths strip.

2. Possible answer: It is easy to find equivalent fractions with denominators of 100 for halves, fourths, and tenths. This allows you to quickly find decimal forms for these fractions. Since $\frac{1}{2}$, $\frac{2}{4}$, $\frac{3}{6}$, $\frac{4}{8}$, $\frac{5}{10}$, and $\frac{6}{12}$ are all equivalent to $\frac{1}{2}$, they all have decimal form 0.5.

3. a. $\frac{1}{6}$ or $\frac{2}{12}$ b. $\frac{3}{10}$ c. $\frac{3}{5}$ or $\frac{6}{10}$ d. $\frac{11}{12}$

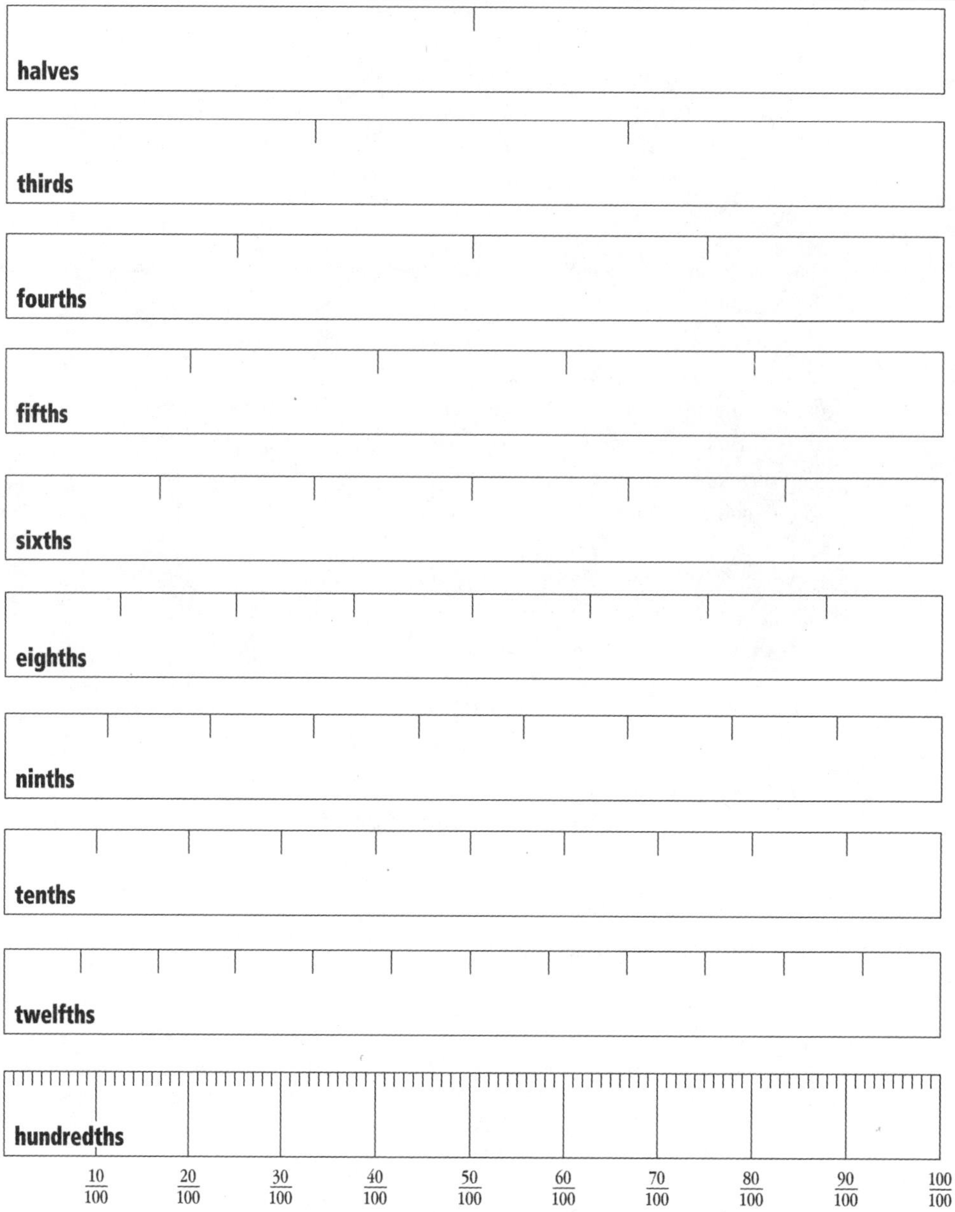

halves

thirds

fourths

fifths

sixths

eighths

ninths

tenths

twelfths

hundredths

$\frac{10}{100}$ $\frac{20}{100}$ $\frac{30}{100}$ $\frac{40}{100}$ $\frac{50}{100}$ $\frac{60}{100}$ $\frac{70}{100}$ $\frac{80}{100}$ $\frac{90}{100}$ $\frac{100}{100}$

Moving from Fractions to Decimals

In 1992, a hurricane swept through the Bahamas, Florida, and Louisiana, destroying many homes and causing lots of damage to land and buildings. The storm was named Hurricane Andrew. Many people lost everything, and had no place to live and very little clothing and food. In response to the disaster, people from all over collected clothing, household items, and food in a relief effort to send to the victims of the hurricane.

One group of students decided to collect food to distribute to some of the families whose homes were destroyed. They would pack what they collected in boxes to send it to the families. The students had to solve some problems while they were packing the boxes.

Problem 5.3

The students had 24 boxes for packing the food they collected. They wanted to share the supplies equally among the families who would receive the boxes. They had small bags and plastic containers to use to repack items for the individual boxes.

The students collected the following items:

48 tins of cocoa mix	6 pounds of Swiss cheese
72 boxes of powdered milk	3 pounds of hot pepper cheese
264 boxes of juice	7 pounds of peanuts
120 boxes of granola bars	5 pounds of popcorn kernels
36 pounds of wheat crackers	475 apples
18 pounds of peanut butter	195 oranges
12 pounds of cheddar cheese	

A. How much of each item should the students include in each box? Explain your reasoning.

B. What operation (+, −, ×, ÷) did you use to find your answers? Why did this operation work?

C. How can your calculator help you decide how to distribute the food items?

■ Problem 5.3 Follow-Up

One student calculated the amount of Swiss cheese to include in each box by entering 6 into her calculator and dividing by 24. Is this a good method? Why or why not?

At a Glance

Launch

■ Tell the story of the students' relief effort.

■ Remind students that explaining their answers is part of solving the problem.

Explore

■ Ask students questions that help them focus on meaning.

Summarize

■ As a class, discuss strategies for determining how to distribute the food items.

■ Talk about how the idea of sharing relates to the division interpretation of fractions.

Answers to Problem 5.3

A. See page 66h.

B. ÷; You are *dividing* the total quantity for each item among the 24 boxes.

C. You can use a calculator to divide the total quantity for each item by 24.

Answer to Problem 5.3 Follow-Up

Possible answer: Yes, because you get 0.25, which is equal to $\frac{1}{4}$, which means each box gets a quarter pound of Swiss cheese. Division is a good way to figure out how to share things equally.

Assignment Choices

ACE questions 25–31 (26–28 require Labsheet 5.2), 33, 34, 40–50, and unassigned choices from earlier problems

Assessment

It is appropriate to use Check-Up 2 after this problem.

Answers

Applications

1. Possible answers: $\frac{25}{100}$ and $\frac{1}{4}$. The decimal value 0.25 *means* $\frac{25}{100}$. Since 25 is $\frac{1}{4}$ of a 100, $\frac{25}{100}$ is equal to $\frac{1}{4}$.

2. Possible answers: $\frac{40}{100}$ and $\frac{2}{5}$. The decimal value 0.40 *means* $\frac{40}{100}$. If you divide 100 into five equal parts, each part would be 20. Two of the five parts would be 40, so $\frac{40}{100}$ equals $\frac{2}{5}$.

3. Possible answer: $\frac{1}{10}$

4. Possible answer: $\frac{4}{10}$

5. Possible answers: $\frac{1}{20}$ or $\frac{1}{25}$

6. Possible answer: $\frac{8}{10}$ or $\frac{4}{5}$

7. $\frac{7}{2}$ or $3\frac{1}{2}$ or 3.5 pounds of pretzels; 4 melons; $\frac{2}{3}$ or about 0.67 pounds of pecans; $\frac{4}{3}$ or $1\frac{1}{3}$ or about 1.33 pounds of peanut butter; $\frac{10}{3}$ or $3\frac{1}{3}$ or about 3.33 pounds of mozzarella cheese; $\frac{2}{9}$ or about 0.22 pounds of grated parmesan cheese; 9 pizza crusts; 6 cans of pizza sauce.

8a. Antonio's fish; Possible explanation: $\frac{2}{3}$ is equivalent to $\frac{16}{24}$, and $\frac{5}{8}$ is equivalent to $\frac{15}{24}$. Antonio's fish is longer because $\frac{16}{24}$ is greater than $\frac{15}{24}$.

8b. Possible answer: $\frac{2}{3}$ is about 0.67, and $\frac{5}{8}$ is about 0.63. It is easy to see that 0.67 is larger than 0.63 since each decimal has the same number of places. Using decimals, it may have been easier for them to tell which fish was longer.

As you work on these ACE questions, use your calculator whenever you need it.

Applications

1. Name three fractions whose decimal equivalent is 0.25. Explain your answer. Draw a picture if it helps explain your thinking.

2. Name two fractions whose decimal equivalent is 0.40. Explain your answer. Draw a picture if it helps explain your thinking.

In 3–6, give a good fraction estimate for the decimal.

3. 0.08 **4.** 0.4 **5.** 0.04 **6.** 0.84

7. Another class collected the following food items to assemble 18 boxes for the hurricane victims:

63 pounds of pretzels	60 pounds of mozzarella cheese
72 melons	4 pounds of grated parmesan cheese
12 pounds of pecans	162 pizza crusts
24 pounds of peanut butter	108 cans of pizza sauce

How much of each kind of food will go into each box?

8. a. Sarah and Antonio went fishing in the Grand River, and each caught one fish. Sarah's fish was $\frac{5}{8}$ of a foot long and Antonio's was $\frac{2}{3}$ of a foot long. Which fish was longer? Explain.

b. If Sarah and Antonio had measured their fish in decimals, would it have been easier for them to tell which fish was longer? Explain.

9. Each small square represents $\frac{1}{100}$. What decimal is represented by this set of grids?

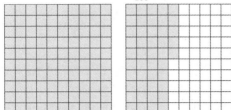

10. Each small square represents $\frac{1}{100}$. What decimal is represented by this set of grids?

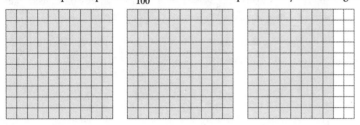

In 11–13, use the fraction strips on Labsheet 5.2 to help you estimate a good decimal equivalent for the fraction.

11. $\frac{3}{4}$ **12.** $\frac{2}{3}$ **13.** $\frac{5}{12}$

In 14–16, find a fraction with a denominator of 10 or less that is a good estimate for the decimal.

14. 0.6   **15.** 0.91   **16.** 0.33

In 17–20, compare the two numbers. Then rewrite the numbers, inserting <, >, or = between them to make a true statement.

17. $\frac{4}{6}$ $\frac{2}{3}$ **18.** 0.34   4

19. $\frac{2}{5}$ $\frac{1}{3}$ **20.** 0.08   0.3

21. Which are easier to compare, fractions or decimals? Why?

9. 1.45
10. 2.80 or 2.8
11. 0.75
12. 0.67
13. 0.42
14. $\frac{3}{5}$
15. $\frac{9}{10}$
16. $\frac{1}{3}$
17. $\frac{4}{6} = \frac{2}{3}$
18. 0.34 < 4
19. $\frac{2}{5} > \frac{1}{3}$
20. 0.08 < 0.3

21. Possible answer: Decimals are usually easier to compare. When fractions have different denominators, it is often difficult to compare them without first finding equivalent fractions.

22. 0.9 > 0.45; Possible explanation: Since 0.9 is the same as 0.90, you can look at the digits after the decimal point—90 and 45—to find which is larger.

23. 0.75 > 0.6; Possible explanation: Since 0.6 is the same as 0.60, you can look at the digits after the decimal point—75 and 60—to find which is larger.

24. 0.6 = 0.60; Possible explanation: Six tenths and sixty hundredths are the same.

25. 0.375; By looking at the fraction strips, the mark for $\frac{3}{8}$ lines up between the marks for 0.37 and 0.38 on the hundredths strip.

26. $\frac{3}{4}$ = 0.75

27. $\frac{5}{8}$ = 0.625

28. $\frac{13}{25}$ = 0.52

29. $\frac{17}{25}$ = 0.68

30. $\frac{1}{20}$ = 0.05

31. $\frac{7}{20}$ = 0.35

32. See page 66i.

33a. $\frac{8}{10}$ or 0.8 of a pizza

33b. See page 66i.

34. Possible answer: When you divide one number by another, you are distributing equal parts of the first number to the number of groups given by the second number. So, you are *sharing* equal parts of the first number to the number of groups given by the second number.

---

22. Which is greater, 0.45 or 0.9? Explain your reasoning. Draw a picture if it helps explain your thinking.

23. Which is greater, seventy-five hundredths or six tenths? Explain your reasoning. Draw a picture if it helps explain your thinking.

24. Which is greater, 0.6 or 0.60? Explain your reasoning. Draw a picture if it helps you explain.

25. James says a fraction is another way to represent a division problem. For example, he says $\frac{3}{8}$ means the same thing as $3 \div 8$. What do you get when you do this division on your calculator? Compare your decimal answer with your fraction strips to see if this is reasonable. Describe your findings.

In 26–31, use your calculator to find a decimal form for the fraction. Then use the hundredths strip or other fraction strips on Labsheet 5.2 to check whether your answer is reasonable.

26. $\frac{3}{4}$ **27.** $\frac{5}{8}$ **28.** $\frac{13}{25}$

29. $\frac{17}{25}$ **30.** $\frac{1}{20}$ **31.** $\frac{7}{10}$

32. Suppose a new student starts school today and your teacher asks you to teach her how to find decimal representations for fractions. What would you tell her? How would you convince the student that your method works?

Connections

33. Ten students went to a pizza parlor together. They ordered eight small pizzas.

a. How much will each student receive if they share the pizzas equally? Express your answer as a fraction and as a decimal.

b. Explain how you thought about the problem. Draw a picture that would convince someone that your answer is correct.

34. Zachary says division should be called a "sharing operation." Why might he say this?

35. If we look through a microscope that makes objects appear ten times larger, 1 centimeter on a metric ruler looks like this:

a. Copy this microscope's view of 1 centimeter. Subdivide the length for 1 centimeter into ten equal parts. What fraction of a centimeter does each of these parts represent?

b. Now think of subdividing one of these smaller parts into ten equal parts. What part of a centimeter does each of the new segments represent?

c. If you were to subdivide one of these new small parts into ten parts again, what part of a centimeter would each of the new small parts represent?

36. Here is what one of the fleas that live on the dog at Mr. Valicenti's fishing camp looks like through the microscope that makes objects appear ten times larger.

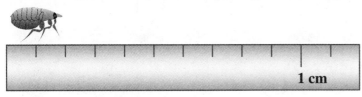

a. About how long is the flea in centimeters?

b. How long would a line of 10 fleas be?

c. How many fleas would it take to equal 1 centimeter?

Connections

35a. See below left.

35b. Each part represents $\frac{1}{100}$ or 0.01 of a cm.

35c. Each part represents $\frac{1}{1000}$ or 0.001 of a cm.

36a. about $\frac{2}{10}$ or 0.2 cm long

36b. about 2 cm long

36c. 5 fleas

35a. Each part represents $\frac{1}{10}$ or 0.1 of a cm.

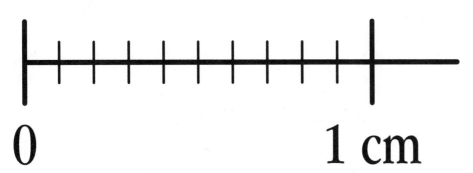

37a. about 1.3 cm long

37b. about 13 cm long

37c. about 65 fleas

37. There are also ferocious flies at Mr. Valicenti's camp.

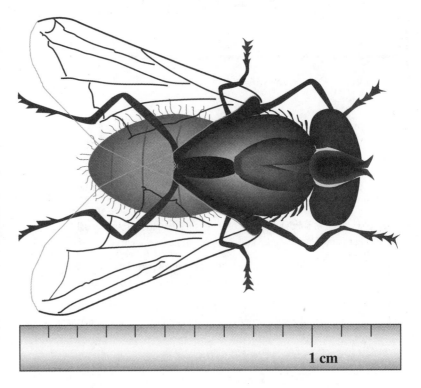

1 cm

a. About how long is the fly in centimeters?

b. If we line up ten of these flies end to end, how long would the line of flies be?

c. About how many fleas lined up end to end (see question 40) would equal the length of a line of ten flies?

38. a. Copy the number line below. Show 0.4 and 0.5 on your number line. Can you place five numbers between 0.4 and 0.5? If yes, place them on your number line with labels. If no, explain why not.

0   1

b. Now, enlarge the line segment from 0.4 to 0.5. Make your new line segment approximately the length of the original number line. Place 0.45 and 0.50 on your new number line. Can you find five numbers that belong between 0.45 and 0.50? If yes, place them on your number line with labels. If no, explain why not.

39. Using the fraction benchmarks 0, $\frac{1}{4}$, $\frac{1}{2}$, $\frac{3}{4}$, and 1 and the number line below, copy and complete the table to show what two fraction benchmarks each decimal is between. Also, tell which benchmark each decimal is nearest.

0          $\frac{1}{4}$ $\frac{1}{2}$ $\frac{3}{4}$ 1

Decimal	Lower benchmark	Upper benchmark	Nearest benchmark
0.17	0	$\frac{1}{4}$	$\frac{1}{4}$
0.034			
0.789			
0.092			
0.9			
0.491			
0.627			
0.36			

38a. See page 66i.

38b. See page 66i.

39. See page 66j.

Extensions

40. See below right.

41. See below right.

42. See page 66j.

43. See page 66j.

44. See page 66j.

Extensions

In 40–44, copy the segment of the number line given. Then, find the "step" by determining the difference from one mark to another. Label the unlabeled marks with decimals. Here is an example:

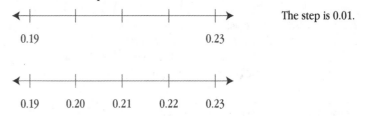

The step is 0.01.

40.    The step is _____ .

0.003   0.403

41.   The step is _____ .

0.198   0.200   0.202

42.    The step is _____ .

0.7634   0.7834   0.8034

43.   The step is _____ .

0.512   0.520

44.    The step is _____ .

0.3   0.4

40. The step is 0.1.

0.003   0.103   0.203   0.303   0.403

41. The step is 0.001.

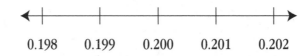

0.198   0.199   0.200   0.201   0.202

In 45–50, find an estimate if you cannot find an exact answer. You may find that making a number line or a diagram is useful in solving the problem. Explain how you reasoned about each problem.

45. What is $\frac{1}{4}$ of 12?

46. What is $\frac{3}{4}$ of 8?

47. What is $\frac{2}{9}$ of 18?

48. What is $\frac{2}{9}$ of 3?

49. What is $\frac{1}{4}$ of 3?

50. What is $\frac{3}{4}$ of 3?

45. See below left.
46. See below left.
47. See page 66j.
48. See page 66j.
49. See page 66k.
50. See page 66k.

Investigation 5: Moving Between Fractions and Decimals 65

---

45. 3; Possible explanation: 3 of 12 equal parts is $\frac{1}{4}$ of the whole.

46. 6; Possible explanation: 6 of 8 equal parts is $\frac{3}{4}$ of the whole.

Possible Answers

1. Dividing the numerator of a fraction by the denominator gives you the decimal equivalent. You can check this by writing the decimal as a fraction with a denominator of 10, 100, 1000, or some other power of ten and then observing that the original fraction and the decimal fraction are equivalent (or close to equivalent).

2. To find a fraction equivalent to a given decimal, use the digits of the decimal as the numerator, and use the place value of the last digit to determine the denominator. For example, the numerator for 0.589 is 589. Since the 9 is in the thousandths place, the denominator is 1000. So 0.589 is equal to $\frac{589}{1000}$.

3. When comparing two decimals, it helps if the decimals are expressed with the same number of decimal places. For example, you can express 0.57 as 0.570 and then easily compare the digits to the right of the decimal point. You can compare digit by digit from the largest place value to the smallest until you find one number larger than the other. Since 570 is greater than 559, 0.57 is larger than 0.559.

Mathematical Reflections

In this investigation, you developed ways to represent fractions as decimals. You used your fraction strips to find fractions and decimals that are close to each other. These questions will help you summarize what you have learned:

1 Describe how to find a decimal equivalent to a given fraction. How can you check your strategy to see that it works?

2 Describe how to find a fraction equivalent to a given decimal. Explain why your strategy works.

3 When comparing two decimals—such as 0.57 and 0.559—how can you decide which decimal represents the larger number?

Think about your answers to these questions, discuss your ideas with other students and your teacher, and then write a summary of your findings in your journal.

Tips for the Linguistically Diverse Classroom

Original Rebus The Original Rebus technique is described in detail in *Getting to Know CMP*. Students make a copy of the text before it is discussed. During discussion, they generate their own rebuses for words they do not understand as the words are made comprehensible through pictures, objects, or demonstrations. Example: Question 3—key words for which students may make rebuses are *comparing* (> or <?), *decimal* (.), *larger* (>).

TEACHING THE INVESTIGATION

5.1 • Choosing the Best

Launch

Problem 5.1 requires students to compare fractions with different denominators. Some of the denominators are larger than those represented on the fraction strips students have used, so they must find new ways to think about equivalent fractions.

There are two ways to pose this problem. You can present it in its current open form and allow students to generate reasonable ways to make comparisons and to decide who should go to the free-throw line. A less open way to pose the problem is to suggest that students consider finding equivalent representations for the three fractions if they think that would be useful for making comparisons.

Tell students the story of the coach's decision. Whichever way you pose the problem, be sure they are thinking about how to compare the girls' performance when each player took and made a different number of free throws in the warm-ups. Students will gain a lot from working in pairs or small groups, as there are different ways to argue for which player should be chosen.

Explore

In earlier investigations, students learned that the numerator and denominator of a fraction can be multiplied by the same number to obtain an equivalent fraction. Some students will quickly convert all three fractions for easy comparison. Other students will need to use concrete representations to help make sense of the problem.

One way to model the problem is by using a hundredths grid, as on Labsheet 5.1, to show each player's performance. This makes converting to a common number of shots (100) easier to visualize.

For the Teacher

If some students are having difficulty dealing with this problem, consider giving them tiles or other manipulatives to represent each player's performance. Tiles of two colors could be used to show each player's data; for example, red tiles could represent making the shot, and yellow tiles could represent missing the shot. Students could build multiple sets of tiles to make 100 trials. For Angela, they would make four sets of her 25 shots—a total of $4 \times 17 = 68$ red tiles and $4 \times 8 = 32$ yellow tiles, which would visually show her average as 68 out of 100.

If some groups finish early, ask them to work on the follow-up questions.

Summarize

A major point that should arise from the summary is that comparisons of situations with different numbers of trials is difficult unless we find equivalent fraction or decimal representations that allow us to make comparisons.

Let students present their arguments. Try to help them distinguish between making decisions based on mathematical evidence and decisions based on irrelevant details such as the names of the players.

Ask questions to help make students' reasoning more precise.

> Angela said she should be chosen because she shot more free throws than the other two players. Carma said she should be chosen because she missed the fewest shots. Emily said that if the coach looked at their trials closely, she would be chosen because although Angela had attempted more shots, she made only two more than Emily, and Carma missed three shots out of ten. What do you think?

Some of your students will probably support each of these positions. Asking them to clarify their assumptions can help the discussion. Are they assuming that the data they have been given will represent performance in the long run for each player? Are they assuming that the conditions of the game will not affect one player more than another?

Make sure that the two main ways of thinking about the problem—by using a hundredths grid to represent the free throws made by each player, and by converting the fractions to equivalent fractions with denominators of 100—are explored in the summary discussion.

You might want to display Transparency 4.2D while the class discusses the hundredths-grid method.

> Can we use a hundredths grids to show the fraction of free throws made by each player? Explain your reasoning.

Reasoning with a hundredths grid will involve shading 68 out of 100 squares for Angela, 75 out of 100 for Emily, and 70 out of 100 for Carma.

> Can we compare the three fractions by first converting them to equivalent fractions with the same denominator?

These fractions can easily be written with 100 as the denominator. Some students might see this quickly, since 100 is 4×25, 5×20, and 10×10.

In one classroom, the teacher gave groups sheets of chart paper on which to record their decisions and their reasoning. Groups displayed their work at the front of the room.

Teacher   Looking at the displays, we can see that not only have you attacked the problem in different ways, but you have reached different conclusions. Which group would like to explain their answer first?

Jordan   We think Angela should shoot the free throw because she made the most. Emily shot almost as many times but made two less.

Mirabel   I disagree. You need to think about what happens in the long run to compare their free-throw attempts. Our group did this by saying, What would happen if each girl shot 100 times and always shot the same way? If each girl shoots 100 times, Angela will make about 68, because it would be like shooting 25 four times and each time she would make 17. Emily will make about 75, because it would be like shooting 5 groups of 20 and for each group she would make about 15. Carma will make about 70, because it will be like 10 groups of 10 and each time she would make about 7. If we look at the players like this—where they have taken the same number of shots—it is obvious Emily should shoot the free throw.

Hearing this argument, several groups decided that they agreed with Mirabel's answer.

Teacher Based on Mirabel's reasoning, several of you now think Emily should go to the free-throw line. Mirabel's group is saying that it is easier to think about this problem by finding a way to compare $\frac{17}{25}$, $\frac{15}{20}$, and $\frac{7}{10}$. They found equivalent fractions with denominators of 100. Why would you choose 100 as denominator to rename the fractions?

Chuck Because they all can be changed to so many 100s.

Teacher How do you know that?

Chuck Because Mirabel showed that 100 was a common multiple for 25, 20, and 10.

5.2 • Writing Fractions as Decimals

Launch

In this problem, students are challenged to find decimal estimates for fractions on the fraction strips with which they are already familiar. The intent is not to build proficiency at using fraction strips to find decimal equivalents, but to use the visual representations to help understand that the same quantity can be represented with two different symbol systems, fractions and decimals. Building this basic visual understanding is important, since the technique of dividing on a calculator—which will be explored shortly—is so efficient that it tends to mask the meaning of fraction-decimal equivalence.

Read the problem to the class. Display a transparency of Labsheet 5.2, and ask the class how they could use the fraction strips to find a good decimal name for $\frac{1}{5}$. This should lead to a visual check between the hundredths strip and the fifths strip or between the tenths strip and the fifths strip.

If students do not suggest it, you might want to demonstrate two methods of finding equivalent decimals. Students can either cut the hundredths strip from the bottom of Labsheet 5.2 so they can position it alongside the other fraction strips, or they can use a straightedge to help sight along the marks on the fraction strips and read off decimal approximations.

Rather than having all groups find decimal equivalents for every fraction strip, you might want to assign two or three strips to each group and have them report their results during the class discussion. You might, for example, divide the task by having two groups responsible for finding the decimal names for the fourths, sixths, and twelfths strips; two responsible for the thirds, eighths, and tenths strips; two responsible for the fifths, sixths, and ninths strips; and two responsible for the thirds, eighths, and twelfths strips.

Groups might write their answers on chart paper or on Transparencies of Labsheet 5.2 for easy sharing with the class. Explain that by the end of the problem, each student should have written the decimal equivalents for all the fraction strips on his or her labsheet. Ask students to keep these labsheets; they will need them for the ACE questions.

Explore

Even though the problem focuses on changing fractions to decimals, visual comparison also allows us to convert from decimals to fractions. As you circulate, ask groups to move both directions using the fraction strips. For example, ask students to find a fraction that is a good estimate for 0.66. They should see that $\frac{8}{12}$, $\frac{6}{9}$, $\frac{4}{6}$, and $\frac{2}{3}$ are all good answers.

Summarize

In the summary, remember that the intent is to build an appreciation that fractions and decimals are different representations of equivalent quantities.

One teacher summarized the problem in the following way. The groups had displayed their findings for the strips for which they were responsible.

Teacher   We all know $\frac{1}{2}$ can be renamed as the decimal 0.5 or 0.50. (*On the board, she wrote $\frac{1}{2}$ = 0.5 or 0.50.*) We can demonstrate this by using the halves and hundredths strips, because $\frac{1}{2}$ easily matches up with one of the lines on the hundredths strip.

What other strips were easy to find decimal names for? (*The class suggested fourths, fifths, tenths, and some of the sixths, eighths, and twelfths fractions.*) Let's start with the easy ones. Which group would like to explain their display for fourths?

One group shared their results. Everyone agreed with their findings, so the class moved on to fifths and tenths. Agreement was again reached. The teacher displayed a transparency of Labsheet 5.2.

Teacher   I want to take a moment to record our results so far. (*She wrote the fraction and decimal name for halves, fourths, fifths, and tenths.*) Could someone explain why $\frac{2}{5}$ and $\frac{4}{10}$ have the same decimal name?

Cheryl   They are equivalent fractions, so they must be the same decimal because decimals are just another way to write fractions.

Teacher   We still have thirds, sixths, eighths, ninths, and twelfths. For which of these was it easiest to find decimal names? (*One student suggested eighths.*) Would the eighths group explain your thinking?

The teacher chose the group who had made this display:

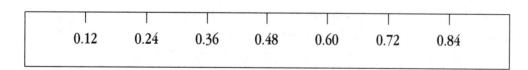

| 0.12 | 0.24 | 0.36 | 0.48 | 0.60 | 0.72 | 0.84 |

Lauren   We started with the $\frac{1}{8}$ mark and lined it up with the 0.12 mark on the hundredths strip. That made a pattern, because $\frac{2}{8}$ is equal to two $\frac{1}{8}$s. We just doubled what we had for $\frac{1}{8}$ and then we just continued in the same way, adding 0.12 each time.

Teacher   What about when you go from $\frac{7}{8}$ to $\frac{8}{8}$? For $\frac{7}{8}$ you have 0.84. If you add 0.12, you would get 0.96, but you know that $\frac{8}{8}$ is 1, not 0.96.

Lauren   Well, that is our one exception, because $\frac{8}{8}$ has to be 1, so it can't be 0.96.

Denis   Your list doesn't make sense to me. I know $\frac{4}{8}$ is the same as $\frac{1}{2}$ and that $\frac{1}{2}$ is the same as 0.5, but your answer says it's 0.48.

Heidi   I agree with Denis. The decimals don't make sense because you have $\frac{2}{8}$ equal to 0.24, and we know $\frac{1}{4}$ and $\frac{2}{8}$ are equivalent fractions and that $\frac{1}{4}$ equals 0.25.

Teacher   Lauren, do you agree that $\frac{2}{8}$ is equal to $\frac{1}{4}$ and that $\frac{1}{4}$ is equal to 0.25? (*Lauren agreed.*) If that is true, what would $\frac{1}{8}$ have to be?

Lauren   It would have to be $0.12\frac{1}{2}$, but you can't have that.

Teacher   I think you could have that amount.

Lauren   Then I want to change some of our others.

Jabe   My group labeled the thirds strip. We want to change our labels now based on what we learned from the eighths group.

Because of this discussion, the class decided to determine very exact decimals for their fractions. The teacher went along with this precision because of level of understanding about what was being communicated.

For the Teacher

The number halfway between 0.12 and 0.13 is actually 0.125. You can help your students to understand this by asking them to picture a thousandths strip. On a thousandths strip, 0.12 would be located at the $\frac{120}{1000}$ mark and 0.13 would be located at the $\frac{130}{1000}$ mark. The number halfway between $\frac{120}{1000}$ and $\frac{130}{1000}$ is $\frac{125}{1000}$, so the number halfway between 0.12 and 0.13 is 0.125.

Ask a few questions that focus on the decimal-to-fraction change.

Use the fraction strips to figure out which fraction is equal to or closest to 0.6? ($\frac{3}{5}$ or $\frac{6}{10}$) How about 0.91? ($\frac{11}{12}$ or $\frac{9}{10}$) 0.33? ($\frac{4}{12}$, $\frac{2}{6}$, or $\frac{1}{3}$) 0.04? (*This lies between 0 and $\frac{1}{12}$, but some students may want to be more precise by making fraction strips with smaller parts, such as halving the twelfths strip to get a twenty-fourths strip.*) 0.84? ($\frac{10}{12}$ or $\frac{5}{6}$)

Students should come away from the summary understanding that they have a visual way to find good estimates in both directions—fractions to decimals and decimals to fractions.

5.3 • Moving from Fractions to Decimals

Launch

This problem is designed to build from students' understanding of the meaning of division to the division interpretation of fractions. It is easy to show students *how* to find a decimal equivalent of a fraction, but it is much harder to help them to understand *why* division is an appropriate interpretation of a fraction.

Tell students the story context, and challenge them to figure out how much of each food item to put into each box. Remind them that explaining how they decided to distribute the goods among the boxes is part of doing the problem.

This problem could be done by first having students work alone and then in pairs to share their thinking. Or, students could work in groups or pairs from the beginning.

Explore

As you circulate, ask questions that focus on meaning.

How many 24s are in 48? What does that tell you about how many tins of cocoa must go in each box?

Which items can be split into fractional parts and which must be in whole units? How does this affect your distribution of supplies?

Summarize

Sharing is a good word to use in discussing the problem, since whole-number division can be used to answer sharing problems and is often used as a context for understanding fractions. Emphasize the connection between sharing a number of items that is evenly divisible by the number of people sharing the items and sharing a number of items that is not evenly divisible by the number of people sharing.

Ask questions that focus on the meaning of division as *equal sharing* to help students understand the division interpretation of fractions. This is a necessary connection for students to understand why dividing with a calculator will change the form of a fraction to a decimal. Have them share their thinking about the problem and the follow-up questions to assess how well they are understanding the relationship between fractions and decimals.

The following discussion took place in one classroom:

Teacher Which food items were easy to share among the 24 boxes, and why?

Angela The first four items: 48 tins of cocoa mix, 72 boxes of powdered milk, 264 boxes of juice, and 120 boxes of granola bars. When you divide each of these by 24, you get a whole number.

Teacher For the cocoa, we can divide 48 by 24 to get 2. What does the 2 mean?

Angela The 2 means that each box will get two tins of cocoa mix.

Teacher What made working with some of the other items more difficult?

Ricardo Some of them didn't turn out to be good amounts, so I didn't know what to do. For the 18 pounds of peanut butter, you can't give every box a pound because there isn't enough.

Frank Yeah, when I tried to divide with my calculator like I did with the others, I pushed 18 ÷ 24 and got 0.75.

Teacher What do you think the 0.75 means? What is a fraction name for 0.75?

Frank Well, 0.75 means 75 out of 100. A fraction name for that is $\frac{3}{4}$. I guess I could give $\frac{3}{4}$ of a pound of peanut butter to each box.

Teacher What do you get when you press 3 ÷ 4 on your calculator? Why does that make sense?

The students at first had a hard time talking about why it made sense to take 3 ÷ 4 and get 0.75, but the more they worked with the numbers and tried to explain what they thought was happening, the more sense they made of the situation.

Ask additional questions to stimulate discussion during the summary.

> While packing the boxes, Lenora said that division should be called a "sharing operation." What could she have meant?
>
> Look at the apples and oranges. Since we don't get a whole number when we divide 475 by 24 or 195 by 24, how will we divide the pieces of fruit among the boxes?

Pull Investigation 5 together by returning to Problem 5.1, which involved comparing fractions with different denominators. That problem can be reworked using a calculator to find decimal representations of the proportion of shots each player made. Ask students to explain how to interpret the decimals. They should recognize that the decimal representation shows the same proportion or fraction, but in an equivalent form based on a power of 10 as the denominator. This is the major objective of this investigation.

Additional Answers

Answers to Problem 5.2

Answers will vary, depending on how carefully students aligned their fraction strips. The answers shown are those obtained by estimating to the nearest tenth and then to the nearest hundredth.

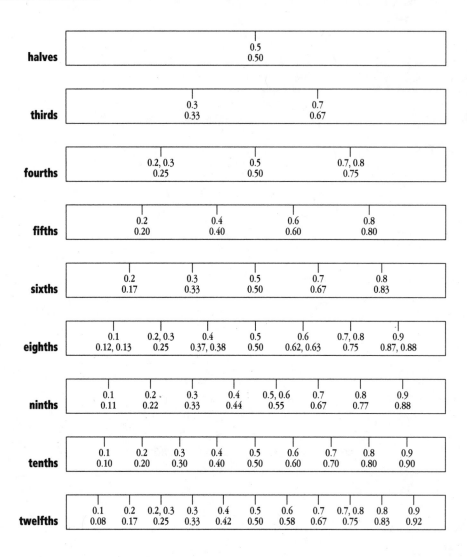

Answers to Problem 5.3

A. tins of cocoa mix: 48 ÷ 24 = 2 tins per box
boxes of powdered milk: 72 ÷ 24 = 3 boxes per box
boxes of juice: 264 ÷ 24 = 11 boxes per box
boxes of granola bars: 120 ÷ 24 = 5 boxes per box
pounds of wheat crackers: 36 ÷ 24 = $1\frac{1}{2}$ or 1.5 pounds per box
pounds of peanut butter: 18 ÷ 24 = $\frac{3}{4}$ or 0.75 pound per box
pounds of cheddar cheese: 12 ÷ 24 = 0.5 or $\frac{1}{2}$ pound per box
pounds of Swiss cheese: 6 ÷ 24 = 0.25 or $\frac{1}{4}$ pound per box
pounds of hot pepper cheese: 3 ÷ 24 = $\frac{1}{8}$ or 0.125 pound per box
pounds of peanuts: 7 ÷ 24 = $\frac{7}{24}$ or about 0.29 pound per box

pounds of popcorn kernels: $5 ÷ 24 = \frac{5}{24}$ or about 0.21 pound per box
apples: $475 ÷ 24 = 19$ apples per box, with 19 apples left over
oranges: $195 ÷ 24 = 8$ oranges per box, with 3 oranges left over

ACE Questions

32. This question is very hard for students at this stage. We expect students to talk about how they can check to see if the division answer makes sense, but not to give a mathematical proof. Possible answer: I would tell the new student to divide the numerator by the denominator on the calculator. I would then tell her to round to the nearest hundredth. To show her that this makes sense, I would write this decimal as a fraction with 100 in the denominator and show her that the original fraction and this fraction are nearly the same amount. I could do this by finding a number to multiply the numerator and the denominator of the original fraction by that gives a denominator close to 100.

Connections

33b. Possible answer: If every student were to receive one piece of every pizza, each pizza would have to be divided into ten equal pieces, with each piece being $\frac{1}{10}$ of a pizza. Each student would receive eight pieces, giving each $\frac{8}{10}$ of a pizza.

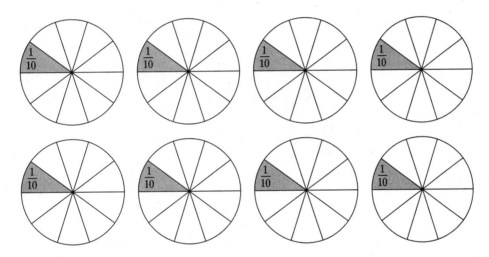

38a. Possible answer:

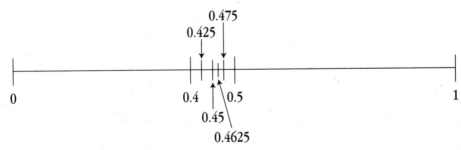

38b. Possible answer:

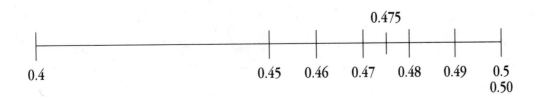

39.

Decimal number	Lower benchmark	Upper benchmark	Nearest benchmark
0.17	0	$\frac{1}{4}$	$\frac{1}{4}$
0.034	0	$\frac{1}{4}$	0
0.789	$\frac{3}{4}$	1	$\frac{3}{4}$
0.092	0	$\frac{1}{4}$	0
0.9	$\frac{3}{4}$	1	1
0.491	$\frac{1}{4}$	$\frac{1}{2}$	$\frac{1}{2}$
0.627	$\frac{1}{2}$	$\frac{3}{4}$	$\frac{3}{4}$
0.36	$\frac{1}{4}$	$\frac{1}{2}$	$\frac{1}{4}$

Extensions

42. The step is 0.01.

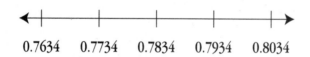

0.7634   0.7734   0.7834   0.7934   0.8034

43. The step is 0.002.

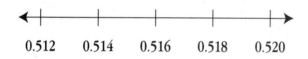

0.512   0.514   0.516   0.518   0.520

44. The step is 0.025.

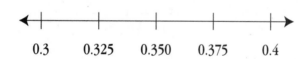

0.3   0.325   0.350   0.375   0.4

47. 4; Possible explanation: 4 of 18 equal parts is $\frac{2}{9}$ of the whole.

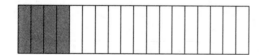

48. $\frac{2}{3}$; Possible explanation: $\frac{2}{9}$ of each of 3 wholes is $\frac{6}{9}$ or $\frac{2}{3}$ of a whole.

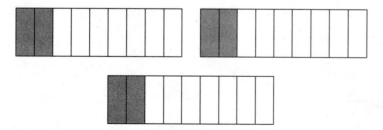

49. $\frac{3}{4}$; Possible explanation: $\frac{1}{4}$ of each of 3 wholes is $\frac{3}{4}$ of a whole.

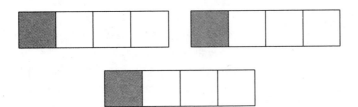

50. $\frac{9}{4}$; $\frac{3}{4}$ of each of 3 wholes is $\frac{9}{4}$ or $2\frac{1}{4}$ wholes.

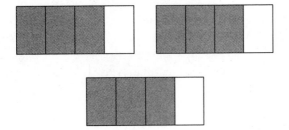

Out of One Hundred

Mathematical and Problem-Solving Goals

- **To use the "out of 100" interpretation of fractions and decimals to develop an understanding of percent**

- **To use the hundredths grid to visualize the concept of percent as meaning "out of 100"**

- **To investigate the relationships among fractions, decimals, and percents and to move flexibly among representations**

- **To understand how to use percent as an expression of frequency, in terms of "out of 100," when a set of data has more or fewer than 100 items**

The context for this investigation is information gathered through surveys. Survey data are most often presented as percents. When we can express data as percents, we can compare different sets of data even when the sample sizes are different.

During students' work in this investigation, we encourage you to focus on two ways to conceptualize percents: using the extension of the visual model of the hundredths grid and the notion of "out of 100." The more readily your students can move among representations of fractions, decimals, and percents, the greater will be their number sense with regard to percent.

In Problem 6.1, It's Raining Cats, students use a database of information about cats to describe the portion of cats who possess some value of an attribute (for example, blue for eye color) as a fraction, a decimal, and a percent. In Problem 6.2, Dealing With Discounts, students consider different ways to express discounts. The goal is to highlight the informal language in daily use and connect it to different representations of quantities. In Problem 6.3, Changing Forms, students move among different forms of representation—sometimes starting with fractions, sometimes decimals, sometimes percents. In Problem 6.4, It's Raining Cats and Dogs, students consider what it means to talk about a percent of a data set involving more than 100 items.

Student Pages	67–83
Teaching the Investigation	83a–84

Materials

For students

- Labsheets 6.1, 6.3, and 6.ACE
- Hundredths strips (from Labsheet 5.2)
- Fraction strips (optional)

For the teacher

- Transparencies 6.1A, 6.1B, 6.2, 6.3, 6.4A, and 6.4B (optional)
- Transparency of newspaper advertisements (optional)

Out of One Hundred

Because fractions that have 100 as their denominator are so useful, there are many ways to represent them. Two ways you have already studied are with decimals and with hundredths grids.

Another useful way to express a fraction with a denominator of 100 is to use a special symbol: the percent symbol, %. **Percent** means "out of 100."

For example, suppose 78 out of 100 middle-school students say they like to swim. You already know how to represent the portion who like to swim with a fraction ($\frac{78}{100}$), a decimal (0.78), and a hundredths grid:

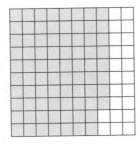

You can also write the fraction of students who like to swim as a percent: 78%.

Look at this grid. How can the shaded part be written as a fraction, a decimal, and a percent?

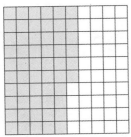

It's Raining Cats

Launch

- Have the class share some of their thoughts about cats.

- With students, examine the cat database and the grid of cat names.

Explore

- Circulate while pairs work with the database, helping them to use logical marking strategies.

- Ask extension questions of pairs that finish early.

Summarize

- Have students share their answers, strategies, and questions they wrote.

- Introduce students to the idea of having more or fewer than 100 items in a database.

Assignment Choices

ACE questions 1–8 (1–7 might require Labsheet 6.1), 12, and unassigned choices from earlier problems

An advertisement in a local newspaper says that a clothing store is having a 30% off sale. This means that all prices have been reduced by $30 for every $100. If a jacket regularly costs $100, it will cost $70 during the sale. You can represent this situation with a hundredths grid:

30% off

You can write the discount on the jacket in other ways as well.
- 30% of the $100 cost = $30 off
- $100 cost – $30 off = $70 sale price

Think about this!

Think of several situations where you have seen or heard percents used. How was percent used in each case?

6.1 It's Raining Cats

A middle-school class has assembled a database of 100 cats owned by students in the school. The database information about each cat is shown on the next three pages.

Cat Database

Cat	Gender	Age (yrs)	Weight (lbs)	Eye color	Pad color
Alex	m	18	11	green	black
Amanda	f	4.5	9.75	blue	gray
Augustus	m	2	10	yellow/green/blue	pink/black
Baguera	m	0.17	13	yellow	brown
Black Foot	m	0.33	1.5	yellow	gray
Blacky	f	1	5	yellow	gray/black
Blue	f	0.25	2	green	gray
Bob	f	4	12	green	black
Boggie	m	3	10	green	pink
Boo	m	3.5	10.75	yellow/green	brown
Boots	m	0.25	3	brown	black
Bosley	m	0.33	1.5	yellow/brown	pink
Bradley	m	0.6	11	yellow	pink/gray
Buffy	m	0.75	8	blue/green	pink
Charcoal	m	11	12	yellow	black
Chelsea	f	2	9	yellow	black
Chessis	f	1.5	6	green	brown
Chubbs	m	1	7	green	pink
Cookie	f	4	9	gold	black
Dana	f	10	8	green	black
Diva	f	3.5	11	green	pink
Duffy	m	1	9	yellow/green	black
Ebony Kahlua	m	1.5	15	blue	brown
Elizabeth	f	10	9	green	pink
Emma	f	4	9.25	gold	pink
Emmie	f	4	7	green	black
Ethel	f	5	8	green	black
Feather	m	2.5	13	green	pink
Fire Smoke	f	0.25	2.5	green/brown	pink
Fluffy	f	5	10	green	pink
Fuzzy	f	1.25	2	green	pink
Gabriel	m	1	7	blue	white
George	m	12	14.5	green	black
Ginger	f	0.2	2	yellow/green	pink
Gizmo	m	4	10	yellow	black
Gracie	f	8	12	green	pink
Gray Kitty	f	3	9	green	gray

Cat	Gender	Age (yrs)	Weight (lbs)	Eye color	Pad color
Grey Boy	m	13	12	yellow	pink
Grey Girl	f	0.2	1.5	gold	pink/black
Grey Poupon	f	5	16	green	pink/black
Hanna	f	3.5	5	yellow	black
Harmony	m	3	12	yellow/green	black
Jinglebob	m	2	18.5	blue	pink
Kali	m	5	16	yellow	black
Kiki	f	1.5	6	green	black
Kitty	m	1.6	10	green	pink
Lady	f	10	8.5	yellow	black
Libby	f	4	8.5	yellow	gray
Lucky	m	4	5	green	pink
Lucy	f	5	10	green	pink
Matilda	f	4.5	9	yellow	pink
Melissa	f	8	11	yellow	pink
Mercedes	f	10	14	green	pink
Midnight	f	10	18	green	pink
Millie	f	10.5	5	blue	black
Miss Muppet	f	11	12	green	pink/black
Mittens	f	14	10.5	yellow	pink
Molly	f	15.5	10	amber	gray
Momma Kat	f	10	6	chartreuse	gray/white
Nancy Blue	f	0.6	5	blue	gray
Newton	m	5	18	yellow	pink
Peanut	f	15	7	green	pink
Peebles	f	5	9	green	black
Pepper	m	2	12	yellow	pink
Pink Lady	f	1.5	6.5	yellow	pink
Pip	m	1	9	yellow	pink
Precious	f	2	12	green	pink
Priscilla	f	3	8.5	green	pink/black
Prissy	f	4	9	green	pink
Ralph	m	3	9	yellow	black
Ravena	f	6	14	yellow	pink/black
Reebo	m	4	12	green	black
Samantha	f	0.2	2	yellow/green	pink
Sassy	f	3	8	yellow/green	gray
Scooter	m	7	16	gold	black
Sebastian	f	3	8	blue	black

Cat	Gender	Age (yrs)	Weight (lbs)	Eye color	Pad color
Seymour	m	0.25	1.5	gold	pink/black
Shiver	m	3	12	yellow/green	pink
Simon	m	0.25	2	green/brown	peach/gray
Skeeter	m	6	13	green	black
Smokey	f	2.5	8	green	black
Smudge	m	8	10	green	gray
Snowy	m	0.5	1.5	gray	gray
Sparky	m	7	12	green	pink
Speedy	m	3	12	blue	pink
Stinky	m	0.17	3.5	yellow	pink
Sweet Pea	f	16	14.5	green	black
Tabby	f	1.5	7	green	black
Tabby Burton	m	1	10	green	black
Terra	m	3	11	green	pink
Thomas	m	4	8	green	pink
Tiger	f	5	13	green	pink
Tigger	f	4	8	yellow	brown
Ting	f	0.25	2.5	green	pink/black
Tom	m	0.25	3	green	gray
Tomadachi	m	1	6.5	gold	pink
Treasure	f	4	8	green	pink
Wally	m	5	10	green	pink/black
Weary	m	8	15	green	pink
Ziggy	f	7	10	gold	pink/black

Did you know?

Ancient Egyptians considered cats to be sacred. Bastet, the Egyptian goddess of love and fertility, was represented as having the head of a cat and the body of a woman. Punishment for harming a cat was severe, and the sentence for killing a cat was usually death. When a cat died, Egyptians shaved their eyebrows as a sign of mourning. Dead cats were often mummified and buried in cat cemeteries.

Tips for the Linguistically Diverse Classroom

Rebus Scenario The Rebus Scenario technique is described in detail in *Getting to Know CMP*. This technique involves sketching rebuses on the chalkboard that correspond to key words in the story or information you present orally. Example: some key words and phrases for which you may need to draw rebuses while discussing the Did you know? feature: *Ancient Egyptian* (a stick figure of an ancient Egyptian), *Baset, the Egyptian goddess* (a figure with the head of a cat and the body of a woman), *sentence for killing a cat* (a stick person next to gallows and a dead cat), *shaved* (razor), *eyebrows* (a face with eyebrows next to the same face without eyebrows), *mourning* (the face without eyebrows crying), *mummified* (a cat wrapped as a mummy), *cat cemeteries* (a tombstone with cat picture).

To help them visualize the different characteristics of the cats in the database, the students made a hundredths grid with each cat's name in one of the squares.

Alex	Boots	Diva	Fuzzy	Hanna	Matilda	Newton	Ravena	Smokey	Thomas
Amanda	Bosley	Duffy	Gabriel	Harmony	Melissa	Peanut	Reebo	Smudge	Tiger
Augustus	Bradley	Ebony Kahlua	George	Jinglebob	Mercedes	Peebles	Samantha	Snowy	Tigger
Baguera	Buffy	Elizabeth	Ginger	Kali	Midnight	Pepper	Sassy	Sparky	Ting
Black Foot	Charcoal	Emma	Gizmo	Kiki	Millie	Pink Lady	Scooter	Speedy	Tom
Blacky	Chelsea	Emmie	Gracie	Kitty	Miss Muppet	Pip	Sebastian	Stinky	Tomadachi
Blue	Chessis	Ethel	Gray Kitty	Lady	Mittens	Precious	Seymour	Sweet Pea	Treasure
Bob	Chubbs	Feather	Grey Boy	Libby	Molly	Priscilla	Shiver	Tabby	Wally
Boggie	Cookie	Fire Smoke	Grey Girl	Lucky	Momma Kat	Prissy	Simon	Tabby Burton	Weary
Boo	Dana	Fluffy	Grey Poupon	Lucy	Nancy Blue	Ralph	Skeeter	Terra	Ziggy

Jane wondered what percent of the cats were female. To answer her question, she used the information in the database and shaded each square in the grid that represented a female cat.

Tang is interested in kittens. He wants to know what percent of the cats in the database are kittens (8 months old or younger) and what percent are adults (over 8 months old).

Answers to Problem 6.1

A. $\frac{54}{100}$, 0.54, 54%

B. $\frac{46}{100}$, 0.46, 46%

C. 54% and 46% add to 100%. Since no cat can be both male and female, the percents must add to a whole.

D. $\frac{17}{100}$, 0.17, 17%

E. $\frac{83}{100}$, 0.83, 83%

F. 17% and 83% add to 100%. No cat can be both a kitten and an adult, so the percents must add to the whole.

Problem 6.1

Using the database and Labsheet 6.1, mark all the cats that are female on one chart and all the cats that are kittens on another chart. When you are finished, answer the following questions.

A. What fraction of the cats are female? Write the fraction as a decimal and a percent.

B. What fraction of the cats are male? Write the fraction as a decimal and a percent.

C. What do you notice about the combined percentage of female and male cats?

D. What fraction of the cats are kittens? Write the fraction as a decimal and a percent.

E. What fraction of the cats are adults? Write the fraction as a decimal and a percent.

F. What do you notice about the combined percentage of kittens and adult cats?

▓ Problem 6.1 Follow-Up

Make up another question that could be answered by looking at the database. Find the answer to your question.

6.2 Dealing with Discounts

Sometimes it is easier to think about a number in one representation than in another. For example, you would probably say that one computer screen is 34% larger than another, rather than saying it is $\frac{17}{50}$ larger than the other.

You will often find it helpful to be able to move among percents, decimals, and fractions. To do this you need to remember that percent means "out of 100," because this can help you to change a percent to a decimal. One way to think about changing representations is to write the percent as a fraction with 100 as its denominator.

When a store offers a discount, such as 20% off, it means that for every $100 an item costs, you get $20 off the price. It also means that for every dollar (100¢) an item costs, you get 20¢ off the price. In other words, you pay $80 for each $100 of the original cost, or, equivalently, 80¢ for each dollar of the original cost.

Investigation 6: Out of One Hundred **73**

Dealing with Discounts

Answer to Problem 6.1 Follow-Up

Possible question: How many cats have a name that starts with the letter B? The answer to this question is $\frac{9}{100}$, 0.09, or 9%.

A pet store is having a big sale.

Problem 6.2

You may want to use fraction strips or hundredths squares to help you to think about these questions.

A. 1. Rewrite the text on the sign for leashes so that the discount is shown as a *fraction off* the original price of the leashes.

2. What will a $10.00 leash cost after the discount?

B. 1. Rewrite the text on the sign for pet carriers so that the discount is shown as a *percent off* the original price of the carriers.

2. What will a $10.00 pet carrier cost after the discount?

C. 1. Rewrite the text on the pet food sign so that the discount is shown as a *percent off* the original price of pet food.

2. Now, write the discount as a *percent* of the original price customers will pay.

3. Rewrite the discount as a *fraction* of the original price customers will pay.

4. Rewrite the discount as a *fraction off* the original price customers will pay.

5. What will $10.00 worth of pet food cost after the discount?

D. 1. Rewrite the text on the pet treats sign so that the discount is shown as a *decimal*.

2. What will $10.00 worth of pet treats cost after the discount?

Answers to Problem 6.2

A. 1. Possible answer: Leashes: $\frac{3}{10}$ off the marked price on the ticket.
   2. $7.00

B. 1. Possible answer: Take 25% off the list price for pet carriers.
   2. $7.50

C. 1. Possible answer: Take 40% off the original price of pet food!
   2. Possible answer: Pay only 60% of the original price for pet food!
   3. Possible answer: Pay only $\frac{3}{5}$ of the original price for pet food!
   4. Possible answer: Take $\frac{2}{5}$ off the original price of pet food!
   5. $6.00

D. 1. Possible answer: All year long you can count on a 0.125 discount on pet treats.
   2. $8.75

Problem 6.2 Follow-Up

1. How can you change a percent to a fraction?

2. How can you change a percent to a decimal?

3. How can you change a decimal to a fraction?

4. How can you change a decimal to a percent?

5. How can you change a fraction to a decimal?

6. How can you change a fraction to a percent?

6.3 Changing Forms

There are many different ways to talk about number relationships. When you are telling a story with data, you have choices about how you express the relationships. Fractions or decimals or percents may be more suitable in certain situations. In this problem you will practice using what you know about changing between fractions, decimals, and percents.

A group of cat owners were asked this question: How much ransom would you be willing to pay if your pet was kidnapped? The table shows how the cat owners responded.

	Percent	Decimal	Fraction
$2000 and up	18%		
From $1500 to $1999		0.03	
From $1000 to $1499			$\frac{3}{100}$
From $500 to $999	25%		
From $1 to $499		0.31	
Nothing			$\frac{1}{5}$

Problem 6.3

Labsheet 6.3 contains the table above and a hundredths grid.

A. Fill in the missing information in your table.

B. Shade in the hundredths grid with different colors or shading styles to show the percent responding to each of the six choices. Add a key to your grid to show what each color or type of shading represents. When you finish, the grid should be completely shaded. Explain why.

Problem 6.3 Follow-Up

1. What percent of cat owners would pay less than $1000 ransom to get their pets back?

2. What percent of cat owners would pay less than $2000 ransom to get their pets back?

Answers to Problem 6.2 Follow-Up

See page 83g.

Answers to Problem 6.3

See page 83h.

Answers to Problem 6.3 Follow-Up

1. 76%

2. 82%

6.3

Changing Forms

At a Glance

Launch

- Talk about the survey and the partially complete table.

- Have groups explore the problem.

Explore

- As you visit the groups, ask them questions about how they are determining their answers.

- Pose extra challenges for groups who finish quickly.

Summarize

- Have a few groups explain their strategies for completing the table.

Assignment Choices

ACE questions 13, 17–20 (17–19 require Labsheet 6.ACE), 22–28, and unassigned choices from earlier problems

It's Raining Cats and Dogs

In a recent survey, 150 dog owners and 200 cat owners were asked what type of food their pets liked. Here are the results of the survey:

At a Glance

Launch

- Work through an example survey with the class, calculating percentages and shading grids.

- Make sure students understand that the surveys in the problem do not involve groups of 100 responses.

Explore

- Circulate among the groups, helping students who are confused about how to reason about the problem.

Summarize

- Have a few groups present their answers and explain their strategies.

- Pose additional questions involving surveys with other than 100 respondents.

Preference	Out of 150 dog owners	Out of 200 cat owners
Human food only	75	36
Pet food only	45	116
Human and pet food	30	48

Problem 6.4

Consider the results of the survey.

A. What kind of food is favored by the greatest number of dogs, according to their owners? Write this number as a fraction, a decimal, and a percent of the 150 dog owners surveyed.

B. What choice is favored by the greatest number of cats, according to their owners? Write this number as a fraction, a decimal, and a percent of the 200 cat owners surveyed.

C. What percent of dog owners reported that their dogs liked either human food only or pet food only? Write this percent as a fraction and a decimal.

D. What percent of cat owners reported that their cats liked either human food only or pet food only? Write this percent as a decimal and a fraction.

■ Problem 6.4 Follow-Up

1. Suppose only 100 dog owners were surveyed, with similar results. Estimate how many would have answered in each of the three categories.

2. Suppose 400 cat owners were surveyed, with similar results. Estimate how many would have answered in each of the three categories.

3. Suppose 50 cat owners were surveyed, with similar results. Estimate how many would have answered in each of the three categories.

Assignment Choices

ACE question 21 and unassigned choices from earlier problems

Assessment

It is appropriate to use the Unit Test after this problem.

Answers to Problem 6.4

A. Human food only; $\frac{75}{150} = \frac{1}{2}$, 0.5, 50%

B. Pet food only; $\frac{116}{200} = \frac{58}{100} = \frac{29}{50}$, 0.58, 58%

C. 80%, 0.8, $\frac{120}{150} = \frac{4}{5}$

D. 76%, 0.76, $\frac{152}{200} = \frac{76}{100} = \frac{19}{25}$

Answers to Problem 6.4 Follow-Up

See page 83h.

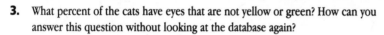

Applications • Connections • Extensions

As you work on these ACE questions, use your calculator whenever you need it.

Applications

In 1–7, use the cat database on pages 69–71 to answer each question. You may want to use the grids on Labsheet 6.1 to help you.

1. What percent of the cats have green eyes (not mixed with green)?

2. What percent of the cats have yellow eyes (not mixed with yellow)?

3. What percent of the cats have eyes that are not yellow or green? How can you answer this question without looking at the database again?

4. What percent of the cats have only pink foot pads?

5. What percent of the cats have only black foot pads?

6. What percent of the cats have only pink/black foot pads?

7. What percent of the cats have foot pads that are not pink, black, or pink/black? How can you answer this question without looking at the database again?

8. 78% of pet owners surveyed say they live in a town where there is a pooper-scooper law in effect.

    a. How would you express this as a decimal?

    b. How would you express this as a fraction?

    c. What percent of people surveyed said they do not live in a town with a pooper-scooper law? Explain your reasoning. Express this percent as a decimal and a fraction.

    d. Can you determine how many people were surveyed? Why or why not?

Answers

Applications

1. 46%

2. 23%

3. 31%; Possible explanation: The percent of cats that have only green eyes or only yellow eyes is 46% + 23% = 69%. So, the percent of cats with eyes that are not only yellow or only green is 100% − 69% = 31%.

4. 41%

5. 28%

6. 10%

7. 21%; Possible explanation: The percent of cats with the three types of foot pads is 41% + 28% + 10% = 79%. So, the percent of cats who have other color foot pads is 100% − 79% = 21%.

8a. 0.78

8b. $\frac{78}{100}$, or $\frac{39}{50}$

8c. 22%, 0.22, $\frac{22}{100} = \frac{11}{50}$; Possible explanation: 100% of the people surveyed represents all the people surveyed, and 78% of those surveyed live in a town with a pooper-scooper law, so 100% − 78% = 22% do not live in a town with a pooper-scooper law.

8d. no; Possible explanation: 78% only tells you *what percentage* of the people surveyed live in a town with a pooper-scooper law; it does not tell you anything about *how many* people were surveyed.

9. 75%; Possible explanation: 100% minus the 25% of the cats weighing less than 7 pounds is 75%. This difference is the percent of cats weighing 7 pounds or more.

10. 28%; Possible explanation: 100% minus the 72% of the cats weighing less than 12 pounds is 28%. This difference is the percent of cats weighing 12 pounds or more.

11. 47%; Possible explanation: The difference between 72% and 25% represents the cats weighing 7 pounds or more and less than 12 pounds.

12. 34%; Possible explanation: 100% represents all the dogs that went to obedience school. The difference between 100% and 66% is the percent of dogs not performing up to par.

13. See below right.

In 9–11, use the following information to answer the questions. Jill noticed that 25% of all the cats in the database on pages 69–71 weigh less than 7 pounds. She also noticed that 72% of all the cats weigh less than 12 pounds.

9. What percent of the cats weigh 7 pounds or more? Explain your reasoning.

10. What percent of the cats weigh 12 pounds or more? Explain your reasoning.

11. What percent of the cats weigh 7 pounds or more and less than 12 pounds? Explain your reasoning.

12. When surveyed, 66% of dog owners who took their dog to obedience school said their dog passed with flying colors. What percent of dog owners said their dogs *didn't* perform up to par? Write an explanation for a friend about how to solve the problem and why your solution works.

13. Copy the table and fill in the missing information.

Percent	Decimal	Fraction
62%		
		$\frac{4}{9}$
	1.23	
		$\frac{12}{15}$

13.

Percent	Decimal	Fraction
62%	0.62	$\frac{62}{100} = \frac{31}{50}$
about 44%	about 0.44	$\frac{4}{9}$
123%	1.23	$\frac{123}{100} = 1\frac{23}{100}$
80%	0.8	$\frac{12}{15}$

In 14 and 15, the large squre represents on whole. Represent the shaded area as a fraction, a decimal, and a percent.

14.

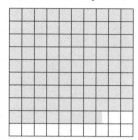

15.

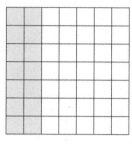

16. Trace the outline of your hand on a hundredths grid. (You can use a copy of Labsheet 4.2.) Estimate what percent of the grid is covered by the area of your hand. Explain how you made your estimate.

In 17–19, use Labsheet 6.ACE to *estimate* what portion of the square is shaded. Explain your reasoning.

17. What percent of this square is shaded? Explain your reasoning.

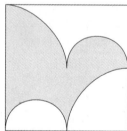

Investigation 6: Out of One Hundred 79

14. $\frac{875}{100}$, 0.875, 87.5%

15. $\frac{14}{49} = \frac{2}{7}$, about 0.286, about 28.6%

16. Answers will vary. Possible explanation: I counted all the squares completely inside the outline and then added parts of the squares inside the outline to make wholes and then added the wholes to my count.

17. about 50%; Possible explanation: If you draw a horizontal line through the center of the square, you can match up the shaded part on the bottom with the unshaded part on the top, and the unshaded part on the bottom with the shaded part on the top, so about half the square is shaded.

18. about $\frac{3}{8}$ or $\frac{2}{5}$; Possible explanations: I covered the square with a grid and counted the shaded part, and the answer was between $\frac{1}{3}$ and $\frac{1}{2}$. I divided the square into four equal squares, and each of the smaller squares had a little less than half shaded.

19. 0.375; Possible explanation: If you divide the figure into four squares of equal size, you can move two of the shaded parts into the upper left square to fill it. Then you have $1\frac{1}{2}$ of the four squares filled, which is 3 half-squares out of 8 half-squares, or $\frac{3}{8}$.

Connections

20. See below right.

18. What fraction of this square is shaded? Explain your reasoning.

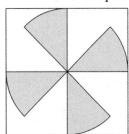

19. What part, in decimal form, of this square is shaded? Explain your reasoning.

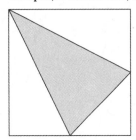

Connections

20. The following percents are a good set of benchmarks to know, because they have nice fraction equivalents and some nice decimal equivalents. The percents are spread out between 0 and 1. Copy the table, and enter the fraction and decimal equivalents for each percent. Use your table until you have learned these relationships.

Percent	10%	$12\frac{1}{2}$%	20%	25%	30%	$33\frac{1}{3}$%	50%	$66\frac{2}{3}$%	75%
Fraction									
Decimal									

20.

Percent	10%	$12\frac{1}{2}$%	20%	25%	30%	$33\frac{1}{3}$%	50%	$66\frac{2}{3}$%	75%
Fraction	$\frac{1}{10}$	$\frac{1}{8}$	$\frac{1}{5}$	$\frac{1}{4}$	$\frac{3}{10}$	$\frac{1}{3}$	$\frac{1}{2}$	$\frac{2}{3}$	$\frac{3}{4}$
Decimal	0.1	0.125	0.2	0.25	0.3	about 0.33	0.5	about 0.67	0.75

Extensions

21. Below are the results when pet owners were asked what kinds of clothing or accessories their pets wear.

	Dogs	Cats
Ribbons	88%	90%
Sweater	65%	20%
Nail Polish	23%	6%
Jeweled Leash	17%	30%
Jewelry	12%	12%

a. Add the percents in the Dogs column. Write the total as a percent, a decimal, and a fraction.

b. Add the percents in the Cats column. Write the total as a percent, a decimal, and a fraction.

c. Why do the columns add to more than 100%?

In 22–25, make a copy of the number line below. Mark and label an approximate point on the number line for each fraction, decimal, and percent given. Use a different number line for each problem.

22. $\frac{3}{8}$, 72%, 1.9, $\frac{4}{3}$

23. 175%, $\frac{7}{9}$, 0.5, 120%

24. 1.35, 0.625, $\frac{8}{5}$, 25%

25. 34%, 0.049, 98%, 1.75

Extensions

21a. 205%; 2.05; $2\frac{5}{100}$

21b. 158%; 1.58; $1\frac{58}{100}$

21c. The columns add to more than 100% because some pets wear more than one item.

22.–25. See page 84.

26. $\frac{1}{2}$, 0.50, 50%

27. $\frac{5}{8}$, 0.625, 62.5%

28. $\frac{11}{60}$, about 0.183, about 18%

In 26–28, determine what fraction is the correct label for the mark halfway between the endpoints of the line segment. Write the fraction as a percent and a decimal.

26.

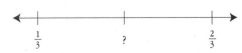

27.

28.

Mathematical Reflections

In this investigation, you explored the relationships among fractions, decimals, and percents. These questions will help you summarize what you have learned:

1. What does *percent* mean?

2. How can you show a percent by using a drawing?

3. a. Describe how you can change a percent to a decimal and to a fraction.

   b. Describe how you can change a fraction to a percent.

   c. Describe how you can change a decimal to a percent.

4. Suppose 12% of students surveyed said they had tried rock climbing. If 100 students were surveyed, how many had tried rock climbing? If 200 students were surveyed? If 150 students were surveyed?

5. A store offers a discount of 30% on all reference books. If a dictionary costs $12.00, what is the amount of the discount? If a book on insect identification costs $15.00, how much will you have to pay for it?

Think about your answers to these questions, discuss your ideas with other students and your teacher, and then write a summary of your findings in your journal.

Possible Answers

1. "out of 100"

2. You can shade part of a hundredths grid to show what part of 100 you want to represent.

3a. For a whole number percent, you make the digits of the percent the numerator of the fraction and 100 the denominator. You find the decimal by dividing 100 into the digits of the percent. For example: $78\% = \frac{78}{100} = 0.78$. If the percent is a fraction, such as 12.5%, this means 12.5 out of 100 or 125 out of 1000. This would give $\frac{125}{1000}$ or 0.125.

3b. If you can change the fraction to an equivalent fraction with 100 as the denominator, the numerator would be the digits of the percent. For example, $\frac{3}{20} = \frac{15}{100} = 15\%$. Or, you can divide the numerator by the denominator to change the fraction to a decimal. Then you write the decimal as a percent. For example, $\frac{3}{4} = 0.75 = 75\%$.

3c. If the decimal is given to the nearest hundredth, the digits of the decimal become the numerator of a fraction with 100 as the denominator. The numerator of the fraction is then the digits of the percent. For example, $0.59 = \frac{59}{100} = 59\%$. If the decimal is something like 0.146, you can write it as 14.6%.

4. 12 of 100; 24 of 200; and 18 of 150

5. The discount is 30% of $12.00, or $3.60. You will pay 70% of $15, or $10.50.

6.1 • It's Raining Cats

Launch

Have a conversation with your class about percents. Ask them what they know about percents and where they have seen or heard percents used. After a few ideas are presented, read and discuss the ideas about the meaning of percent that are presented on page 67 of the student edition.

Begin a class discussion about cats, allowing students to share thoughts about pets they have. Then introduce students to the cat database, perhaps displaying the portion of the database shown on Transparency 6.1B. Ask students what kinds of information are given in the database, and focus their attention on specific information.

> What can you tell me about Fire Smoke? Is Fire Smoke a big cat, a medium-size cat, or a small cat? (*small*) How do you know? (*She is only 0.25 years old and weighs only 2.5 pounds.*)

> The table says that Fire Smoke is 0.25 years old. How old is this in months? (*3*) How did you figure this out? (*0.25 years is $\frac{1}{4}$ of a year. A year is 12 months, so $\frac{1}{4}$ of a year is 3 months.*)

> A kitten is a cat that is 8 months old or younger. Is Fire Smoke a kitten or an adult cat? (*a kitten.*)

Focus students' attention on Labsheet 6.1. Have students work in pairs on this problem.

> Use the grid to keep track of your counting. One person in your pair can shade, color, or label one copy of the grid for sex of the cats, and the other person can mark another grid for adult or kitten. Use any marking scheme that makes sense to you. Express your answers as fractions, decimals, and percents.

Explore

If you notice students having trouble keeping track of what they are marking, you may want to have them compare their charts with another pair's chart before they compute the fractions, decimals, and percents. The notion of checking their data is an important habit for students to develop.

Groups that finish quickly can move on to the follow-up, which asks them to use the database to write another question about the cats. For example, writing questions about weight would require making decisions about what intervals to use to categorize the data set.

Summarize

Allow students to share the fractions, decimals, and percents they found and how they found their answers. One concept you want conveyed is that the fractions, decimals, and percents that describe nonoverlapping categories and that cover the full range of possibilities must add to 1 or 100%. For example, the sum of adult cats and kittens must be 100%, because all cats are in one of these two categories and no cat is in both.

If any pairs have written and explored an additional interesting question, let them share what they found. They should explain their question, what data they needed to tally, and how they found the answer.

Ask questions to better focus students on what they are learning.

Why is it easy to move among fractions, decimals, and percents in this database?

Moving among different representations is easy because the number of cats is 100, and a simple count will give the fraction of cats having the characteristic being considered. Since this fraction will have a denominator of 100, changing to decimal form is easy. Moving from a decimal in hundredths to a percent is also simple because percent means "out of 100."

Suppose we had 150 cats in a database and the same fraction of cats were female. How many would be female?

Students may see different ways to reason about this. In their work, they found that 54% of the 100 cats are female. They might think of the 150 as being 100 plus half of 100. This means that 54 plus half of 54 cats would be female, or 81 out of 150.

Suppose we had only 50 cats, but the percent of females was still 54%. How many are female?

Here we have half of 100 cats, so half of 54 are female, which is 27.

While this strategy is not as efficient as multiplying 0.54×150 or 0.54×50, students need to reason with informal strategies before they move to formal, more efficient strategies. We return to the theme of greater than 100 in a later problem; students need not fully understand the concept now. Just raise questions to get students thinking from the start about what "out of 100" means when the base is not 100. How can we think about scaling up or down from 100 to fit a particular situation?

6.2 • Dealing with Discounts

Launch

This problem looks at different forms of advertising to help students become skillful at interpreting what the information in an ad means in terms of fractions, decimals, and percents.

A good way to launch the activity is to put the following advertisement (or a similar one from your local paper) on the overhead.

Here is a typical advertisement you might see in a store or in the newspaper.

Turtleneck Shirts!

An extra 20% discount will be taken off the ticket price at the register

What would the sign say if the store decided to give the discount as a fraction off the ticket price?

Students will have to find a fraction representation for 20%. One way to reason about this is to first find a decimal representation, 0.20, and then a fraction, $\frac{20}{100}$. Others may write $\frac{20}{100}$ right away and explain that since percent means "out of 100," if you have 20% you must have 20 out of 100.

Is the store likely to use the fraction $\frac{20}{100}$, or will it use an equivalent fraction with a smaller denominator? (*The store would be more likely to use $\frac{1}{5}$.*)

When you feel students are ready, explain that they are to rewrite the signs in the problem in the ways specified. In each problem, they will also figure out the price of $10.00 worth of merchandise after the discount.

Explore

As you visit the groups, listen to how students are thinking about the problem. Look for good strategies that should be shared with the class during the summary. Ask questions to help students tie their reasoning to what they already know about fractions, decimals, and percents. Remind them to revisit pages 67 and 68 to look again at what *percent* and *percent off* mean. Encourage students to make drawings if they would find them helpful.

Students may struggle with changing $12\frac{1}{2}$% to a decimal in question D. Ask them how they would write 12% as a decimal and how they would write 13% as a decimal, which exposes the fact that $12\frac{1}{2}$% is between 12% and 13% and means the decimal is between 0.12 and 0.13. On each dollar, the discount will be 12.5 cents. This means that for $10.00, the discount will be $1.25. Students may need to shade a grid to understand $12\frac{1}{2}$%.

Summarize

Have students present their answers and carefully explain how they thought about the problems. You may want to expand the conversation by asking students to find discounts on prices other than $10.00. You might want to revisit the problem you used in the launch.

The first ad we talked about offered 20% off turtleneck shirts. Suppose you visit this store and look at the ticket on a turtleneck, and this is what you see:

```
┌─────────────────────────────┬──────────┐
│    Style Number 60275       │          │
│  ○ Dept. 27                 │  $19.00  │
│    Size M                   │          │
└─────────────────────────────┴──────────┘
```

What would the price be after the discount is taken at the register?

Students will find several ways to reason about this. They may realize that this means 20¢ off each dollar; for $19, this is 19 × 20 which gives 380 cents or $3.80 off the $19.00. So the price is $15.20. Other students might see that this will be 80¢ for each of the 19 dollars and multiply 80 × 19 to get 1520 cents or $15.20.

Discuss the conversions in the follow-up questions.

6.3 • Changing Forms

Launch

In this problem, students think further about the relationships among percents, decimals, and fractions. We are not looking for the emergence of established algorithms; students will be best served by having time to make sense of the questions and to devise their own ways of thinking about them.

Tell students about the survey data that was gathered. Pass out a copy of Labsheet 6.3 to each student. You may want to display Transparency 6.3.

Pet owners were asked what ransom they would be willing to pay to get a pet back from a kidnapper. This table shows the data organized in a slightly strange way. We have to fill in the missing parts to be able to compare responses across the money categories.

You have already changed some percents to decimals and fractions, and some decimals to fractions and percents. Use what you already know to complete the table. Notice that you are asked to shade the grid to represent the situation and to make observations about your grid when you are finished.

Explore

As you circulate, help students troubleshoot. You can do this by asking questions:

> How do you find an equivalent fraction?
>
> What benchmark is the number near?
>
> Is your answer reasonable?

Pose further challenges for groups who finish early.

> Estimate how many people would respond with "$2000 and up" if a total of 1000 people had been surveyed. What if only 300 people had been surveyed?

Summarize

Have students talk about their answers and their strategies for finding solutions. Here are some strategies students have used:

■ To change a percent to a fraction, make the number in front of the percent sign the numerator, and make the denominator 100.

■ To change a percent to a decimal, first check whether the percent is a one- or two-digit number. If the percent is a two-digit number, drop the percent sign and put a decimal point in front of the first digit. If the percent is a one-digit number, like 5%, drop the percent sign but make sure you write your number so that as a decimal it means "out of 100"—here you would have to write 0.05.

■ To change a decimal to a fraction, make the decimal number the numerator of the fraction, and make the denominator whatever place value the last digit of the decimal is in. For 0.32, the numerator would be 32, and the denominator would be 100 because the 2 is in the hundredths place.

■ To change a two-digit decimal to a percent, just get rid of the decimal point and write the number and a percent sign. For 0.31, you just write 31%. For a decimal like 0.2, write an equivalent decimal with a denominator of 100—0.20—then write it as a percent, 20%.

■ If the denominator of a fraction is a factor of 10 or 100, just write an equivalent fraction with a denominator of 10 or 100. Whatever the new numerator is will be the decimal preceded by a decimal point. If the fraction has a denominator like 8 or 3, you either memorize the relationships or use a calculator to divide the denominator into the numerator, which will give you the decimal equivalent.

■ Changing a fraction to a percent is easier if you can first change it to a decimal, as it's easy to change decimals to percents. You might also be able to use benchmarks to estimate the percent a fraction is near.

6.4 • It's Raining Cats and Dogs

Launch

This problem focuses on percents of numbers greater than 100. A good way to launch the problem is to work through an example with 100 first as a way to help students understand what they need to think about to solve the problem. You might want to use the example on Transparency 6.4A.

In a recent survey, 100 dog owners and 100 cat owners were asked what type of food their pets liked. Here are the results of the survey.

Preference	Out of 100 dog owners	Out of 100 cat owners
Human food only	50	27
Pet food only	30	58
Human and pet food	20	15

How do the preferences for dogs compare with the preferences for cats?

You want students to recognize that percents are good ways to make comparisons. They could, for example, easily compare the dogs' preference for human food to the cats' (50%, 0.50, $\frac{50}{100}$ for dogs; 27%, 0.27, $\frac{27}{100}$ for cats) and their preferences for pet food (30%, 0.30, $\frac{30}{100}$ for dogs; 58%, 0.58, $\frac{58}{100}$ for cats).

What percent of dog owners report that their dogs either liked human food only or dog food only? *(50% + 30% = 80%)* Can you write your answer in a different form? *(0.80, $\frac{80}{100}$)*

Shading a hundredths grid can reinforce how a pictorial representation helps to visualize comparisons. Here are possible ways to shade the two grids for each of the three categories.

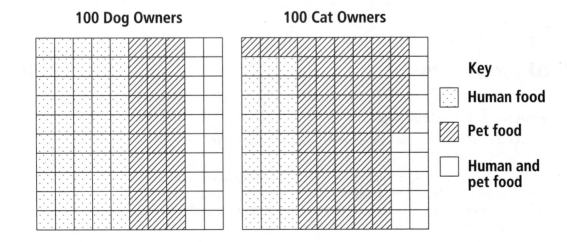

100 Dog Owners   **100 Cat Owners**

Key
- Human food
- Pet food
- Human and pet food

If your students are struggling, make up another problem with a data set of 200 and work through it as a class.

When your students are ready, read the survey results on page 76 of the student edition. Make sure students realize that this time the results are not based on samples of 100. Put students into pairs or small groups, and encourage them to record the ways they are thinking about the problem.

Explore

Pay attention to students' strategies as you visit the groups. Some students will see that 75 out of 150 is the same as 150 out of 300, which they recognize as $\frac{1}{2}$. Others will divide 75 by 150 on their calculators and get 0.5. For students who say that 75 out of 150 means 75%, talk with them about what *percent* means and refer them to page 67 in the student edition.

Some students will look at 45 out of 150 and reason that we should look at 150 in three batches of 50 and look at 45 in three batches of 15. This means that we have $\frac{15}{50}$, which is $\frac{30}{100}$ or 30%.

For the 200 cat owners surveyed, students may reason in hundreds by splitting the 200 into two groups and splitting the amount in each category into two groups. Some students may need to shade two hundredths grids to help think about this. Allow students to stay at this concrete level of thinking if they need to; don't rush to show them that division will immediately give the related decimal.

Summarize

Have groups report their answers and share their strategies. Use this opportunity to help them build flexible ways to think about finding percents when the total is not 100.

> Suppose the person conducting such a survey in your neighborhood only surveyed 75 dog owners. If, of the 75 people who responded, 24 reported that their dogs preferred human food, what percent is this?

Though the total is less than 100, students can scale up to 100. Students might reason that 75 is 3 of 4 equal parts of 100, and if we divide 24 into 3 equal parts we get 8 in each part. Therefore, if we scale up to four parts, we have 32 people reporting that their dogs prefer human food, which is 32%.

Additional Answers

Answers to Problem 6.2 Follow Up

1. Possible answer: Make the number in front of the percent sign the numerator, and make the denominator 100.

2. Possible answer: If the percent is a two-digit number, drop the percent sign and put a decimal point in front of the first digit. If the percent is a one-digit number, like 5%, drop the percent sign but make sure you write your number so that as a decimal it means "5 out of 100"—here you would have to write 0.05.

3. Possible answer: Make the decimal number the numerator of the fraction, and make the denominator whatever place value the last digit of the decimal is in. For 0.32, the

numerator would be 32, and the denominator would be 100 because the 2 is in the hundredths place.

4. Possible answer: With a two-digit decimal, just write the number and a percent sign. For 0.31, you just write 31%. For a decimal like 0.2, write an equivalent decimal with a denominator of 100—0.20—then write it as a percent, 20%.

5. Possible answer: If the denominator is a factor of 10 or 100, just write an equivalent fraction with a denominator of 10 or 100. Whatever the new numerator is will be the decimal preceded by a decimal point. If the fraction has a denominator like 8 or 3, you either memorize the relationships or use a calculator to divide the denominator into the numerator, which will give you the decimal equivalent.

6. Possible answer: Changing a fraction to a percent is easier if you can first change it to a decimal, as it's easy to change decimals to percents. You might also be able to use benchmarks to estimate the percent a fraction is near.

Answers to Problem 6.3

A.

	Percent	Decimal	Fraction
$2000 and up	18%	0.18	$\frac{18}{100}$ or $\frac{9}{50}$
From $1500 to $1999	3%	0.03	$\frac{3}{100}$
From $1000 to $1499	3%	0.03	$\frac{3}{100}$
From $500 to $999	25%	0.25	$\frac{25}{100}$ or $\frac{1}{4}$
From $1 to $499	31%	0.31	$\frac{31}{100}$
Nothing	20%	0.20	$\frac{20}{100}$ or $\frac{1}{5}$

B. The total percentage is 100% because the results of the survey are reported for all the people who responded; that is, 100%. We don't know the number of people who responded, but we do know that the "whole" is 100%.

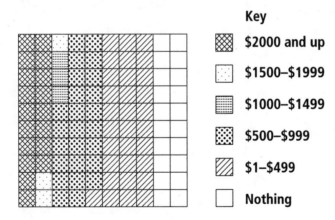

Key

▨ $2000 and up

▦ $1500–$1999

▦ $1000–$1499

▧ $500–$999

▨ $1–$499

☐ Nothing

Answers to Problem 6.4 Follow-Up

1. **Preference**   **Out of 100 dog owners**

   Human food only   50

   Pet food only   30

   Human and pet food   20

2. **Preference**   **Out of 400 cat owners**

   Human food only   72

   Pet food only   232

   Human and pet food   96

3. | Preference | Out of 50 cat owners |
   |---|---|
   | Human food only | 9 |
   | Pet food only | 29 |
   | Human and pet food | 12 |

ACE Questions

Extensions

22.

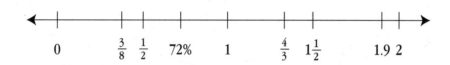

23.

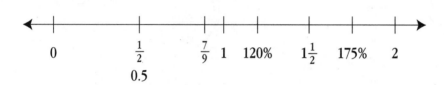

24.

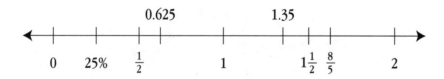

25.

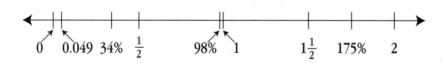

Assessment
Resources

Check-Up 1

1. This sketch shows part of a ruler. The main marks indicate inches.

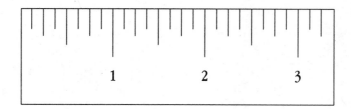

   How do you think each of the marks between the inches should be labeled? Explain your answer.

2. Use your fraction strips or another method to compare the two fractions in each pair. Insert the correct sign:
   <, >, or =.

   a. $\frac{8}{12}$ $\frac{3}{4}$ b. $\frac{5}{8}$ $\frac{6}{10}$ c. $\frac{2}{3}$ $\frac{5}{6}$ d. $\frac{2}{4}$ $\frac{7}{12}$ e. $\frac{3}{8}$ $\frac{3}{12}$

3. Find three different fractions between the benchmarks $\frac{1}{2}$ and $\frac{3}{4}$.

Check-Up 1

4. Julie's math class and Dave's math class are selling sub sandwiches as a fund-raiser. Each class has a goal of $150. Julie said her class was closer to the goal than Dave's class because her class had earned $\frac{2}{3}$ of their goal, and Dave's class had earned $\frac{5}{8}$ of their goal. Was Julie right? Explain your answer.

5. Estimate and mark where the number 1 will be on each number line. The length that represents the whole may be different on each number line.

   a.

   b.

   c.

6. Order these numbers from smallest to largest:

   $1\frac{7}{10}$ $1\frac{15}{18}$ $\frac{24}{15}$

Quiz

1. Give a fraction name for the shaded part of the figure below. Explain how you figured out what fractional part of the whole was shaded.

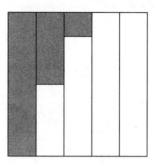

2. Lynette found a worm that is $\frac{2}{3}$ of the length of your fraction strip. How many worms exactly like hers would you need to put end to end to equal two times the length of your fraction strip? Explain your answer.

3. Decide whether each of the following statements is true or false, and explain why:

   a. If you compare two fractions with the same denominator, the fraction with the greater numerator is greater.

   b. If you compare two fractions with the same numerator, the fraction with the greater denominator is greater.

Quiz

4. Antonio's father agreed to help with an Events Night at camp by setting up a pizza stand to sell pizza to visiting family and friends. Antonio's father has only one size of pizza pan, a circular pan with a 16-inch diameter. The class committee decided they wanted him to sell pizza by the slice and to sell small slices and large slices. Antonio's father has a cutting form that can cut a pizza into 12 slices and another form that can cut a pizza into 8 slices.

   a. If a family bought three small slices and three large slices, what fraction of a pizza did they buy? (You might want to draw a picture to help you.)

   b. How much more pizza (what fraction) would they need to buy to purchase a whole pizza?

   c. How many different ways can you combine small slices and large slices to make a whole pizza? Write each of your responses as number sentences. For example: $\frac{2}{8} + \frac{9}{12} = 1$ means that two large slices and nine small slices will make one whole pizza.

5. In each pair of pencils, the length of the new pencil is about what fraction of the length of the old pencil?

   a.

   b.

Name _____ Date _____

1. For each figure below, give a fraction name and a decimal name for the shaded part.

    a.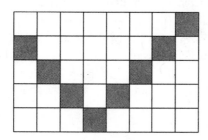

    b.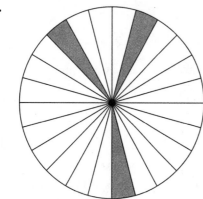

2. Arrange these decimals in order from smallest to largest:

    6.00   0.6   0.006   0.60   0.06   0.00006

3. On each figure below, shade the indicated decimal amount.

    a.   0.375

    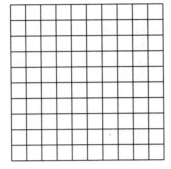

    b.   0.6

    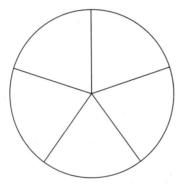

    c.   0.05

    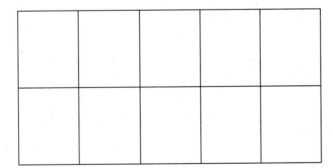

4. Rename each of the decimal amounts in question 3 with fraction names.

    a. _____   b. _____   c. _____

Check-Up 2

5. On the strip below, mark and label where each of these decimals is: 0.09, 0.9, 0.19, 0.190, 0.019.

$\frac{1}{10}$ $\frac{2}{10}$ $\frac{3}{10}$ $\frac{4}{10}$ $\frac{5}{10}$ $\frac{6}{10}$ $\frac{7}{10}$ $\frac{8}{10}$ $\frac{9}{10}$ $\frac{10}{10}$

6. For each number line, fill in the missing decimal numbers. For example, filling in the missing decimal numbers on this number line:

0.019   0.023

would give you this:

0.019   0.020   0.021   0.022   0.023

a.

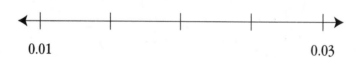

0.01   0.03

b.

0.3061   0.3161   0.3261

c.

0.4302   0.6302

Assign these questions as additional homework, or use them as review, quiz, or test questions.

1. This is a carnival game that tests strength. The player hits the block with a mallet and the force of the blow sends a metal ringer up the pole. If the player uses enough force, the ringer rings the bell at the top of the pole and the player receives the top prize of 100 points. The player receives fewer points for hits that only send the metal ringer partway up the pole. The points can be traded for tickets to rides at the carnival.

   a. Where should marks be made on the pole for each of the game point amounts? Mark them on the pole.

10 points	25 points	35 points
70 points	85 points	100 points

   b. What fraction of the pole would each of the marks in part a represent?

   c. What payoff, in game points, should be given for sending the metal ringer $\frac{1}{3}$ of the way up the pole?

   d. What payoff, in game points, should be given for sending the ringer $\frac{3}{5}$ of the way up the pole?

   e. What payoff, in game points, should be given for sending the ringer $\frac{2}{8}$ of the way up the pole?

   f. What payoff, in game points, should be given for sending the ringer $\frac{3}{4}$ of the way up the pole?

   g. Miki's hit sent the metal ringer $\frac{5}{8}$ of the way up the pole. Taylor's hit went $\frac{6}{9}$ of the way to the top. Who received the most game points? Why?

2. You are invited to go out for pizza with several friends. When you get there, your friends are sitting in two separate groups. You can join either group. If you join the first group, there will be a total of 4 people in the group and you will be sharing 6 small pizzas. If you join the second group, there will be a total of 6 people in the group and you will be sharing 8 small pizzas. If pizza will be shared equally in each group—and you are *very* hungry—which would you rather join? Explain your choice.

3. a. Three is what fractional part of 12?

    b. Five is what fractional part of 20?

    c. Two is what fractional part of 9?

    d. Seven is what fractional part of 17?

4. Samuel is getting a snack for himself and his little brother. There are two muffins in the refrigerator. Samuel takes half of one muffin for himself and half of the other muffin for his little brother. His little brother complains that Samuel got more. Samuel says that he got $\frac{1}{2}$ and his brother got $\frac{1}{2}$. What might be the problem?

5. Your best friend was absent when your class learned how to compare decimal numbers. Write a set of directions that would help your friend understand how to compare decimal numbers.

6. The following prices were posted at a local store.

    What is wrong with these signs?

7. Use these numbers to fill in the blanks so that the story makes numerical sense:

645    $\frac{3}{4}$ 65   215   75   161.25   35 $\frac{1}{4}$ 330.65

Events Night a Huge Success!

The Events Night held by Mr. Martinez's and Ms. Swanson's middle-school classes was a success, raising a total of $_____. The teachers estimated the large turnout of middle-school students included over_____ of the building's student population. Over half of the money, $_____, was earned by the food booths. _____ game tickets were sold, raising $_____, which represented_____ of the money. The tickets were_____ cents each. Most of the money that was raised, _____%, will go toward paying for the class camping trip, and the other _____% will be used to pay expenses.

8. In each of the sets of numbers below, one number is not equivalent to the others. Tell which one is not like the others and explain why.

a. 0.60   0.6   6%

b. $\frac{1}{25}$ 25%   0.25

c. 0.75   34% $\frac{3}{4}$

9. Write a benchmark fraction that is close to each of these percentages:

a. 23.6%   b. 45.4545%

Name _____ Date _____

1. In each figure, express the area shaded and the area not shaded as percents.

   a. 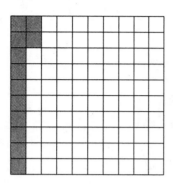   **b.** 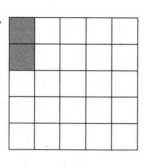   **c.**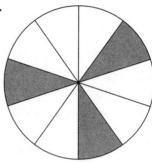

   % shaded _____   % shaded _____   % shaded _____

   % not shaded _____   % not shaded _____   % not shaded _____

2. Write each of the following as a fraction, decimal, and percent.

   a. 30 days out of 100 days   **b.** 55¢ compared to 100¢

   c. 20 correct out of 25 problems   **d.** 3 out of 4 games won

   e. 21 mountain bikes out of 40 bikes   **f.** 5 misspelled words out of 30 words

3. At the pizza shop, 16-inch-diameter pizzas sell for $9.85. The shop has decided to sell pizza by the slice. One cutting form can cut a pizza into 12 slices, and another form can cut a pizza into 8 slices.

   a. To make at least $9.85 on each whole pizza sold, how much should the shop charge for a large slice? (Try to come as close to $9.85 as you can, but make your price easy to handle in terms of making change.)

Unit Test: In-Class Portion

b. To make at least $9.85 on each whole pizza sold, how much should the shop charge for a small slice? (Try to come as close to $9.85 as you can, but make your price easy to handle.)

c. Explain why you think your answers for parts a and b are appropriate.

4. In a recent survey of 600 people, 20% said chocolate chip cookie was their favorite ice cream. How many people in the survey favored chocolate chip cookie ice cream? Explain your answer.

5. Of the people in your math class today (including your teacher), what percent are male? _____

What percent are female? _____

Unit Test: Individual Research

Find two different articles in newspapers or magazines that contain fractions, decimals, or percents. If one article uses mainly one of these forms, the other article must contain at least one of the other two forms.

Write a one- to two-paragraph summary of each article. In your explanation, tell how rational numbers were used in the article and what they represent. Turn in each article with your explanation attached.

Name _____ Date _____

Journal Organization

_____ Problems and Mathematical Reflections are labeled and dated.

_____ Work is neat and easy to find and follow.

Vocabulary

_____ All words are listed.   _____ All words are defined and described.

Quizzes and Check-Ups

_____ Quiz   _____ Check-Up 1

           _____ Check-Up 2

Homework Assignments

_____ _____

_____ _____

_____ _____

_____ _____

_____ _____

_____ _____

_____ _____

_____ _____

_____ _____

_____ _____

_____ _____

_____ _____

_____ _____

Self-Assessment

Vocabulary

Of the vocabulary words I defined or described in my journal, the word _____ best demonstrates my ability to give a clear definition or description.

Of the vocabulary words I defined or described in my journal, the word _____ best demonstrates my ability to use an example to help explain or describe an idea.

Mathematical Ideas

1. **a.** I learned these things about fractions, decimals, and percents in *Bits and Pieces I*:

   b. Here are page numbers of journal entries that give evidence of what I have learned, along with descriptions of what each entry shows:

2. **a.** These are the mathematical ideas I am still struggling with:

   b. This is why I think these ideas are difficult for me:

   c. Here are page numbers of journal entries that give evidence of what I am struggling with, along with descriptions of what each entry shows:

Class Participation

I contributed to the classroom discussion and understanding of *Bits and Pieces I* when I . . .
(Give examples.)

Answer Keys

Answers to Check-Up 1

1. Possible answer: Each space represents an eighth, because there are eight spaces between each number. Each mark would increase by an eighth: $\frac{1}{8}, \frac{2}{8}, \frac{3}{8}, \frac{4}{8}, \frac{5}{8}, \frac{6}{8}, \frac{7}{8}$. If the fractions are reduced, the marks would be $\frac{1}{8}, \frac{1}{4}, \frac{3}{8}, \frac{1}{2}, \frac{5}{8}, \frac{3}{4}, \frac{7}{8}$.

2. a. $\frac{8}{12} < \frac{3}{4}$ b. $\frac{5}{8} > \frac{6}{10}$ c. $\frac{2}{3} < \frac{5}{6}$ d. $\frac{2}{4} < \frac{7}{12}$ e. $\frac{3}{8} < \frac{3}{12}$

3. Possible answers: $\frac{2}{3}, \frac{53}{100}, \frac{6}{10}, \frac{712}{1000}$

4. Yes, Julie was right. Possible explanations: Dave's class had earned $\frac{5}{8}$ ($150) = $93.75, and Julie's class had earned $\frac{2}{3}$ ($150) = $100. Also, $\frac{5}{8} = \frac{15}{24}$ and $\frac{2}{3} = \frac{16}{24}$, so $\frac{5}{8} < \frac{2}{3}$.

5. a.

b.

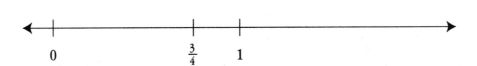

c.

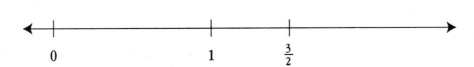

6. $\frac{24}{15}$ $1\frac{7}{10}$ $1\frac{15}{18}$

Answers to Quiz

1. The portion of the figure that is shaded is $\frac{10}{30}$ or $\frac{1}{3}$. If you divide the rectangle into 30 equal pieces, 10 of them are shaded, and $\frac{10}{30}$ equals $\frac{1}{3}$.

2. 3 worms; $3(\frac{2}{3}) = 2$. Students will probably draw a picture to show this, which is quite acceptable at this time.

3. a. true; Possible explanation: If the denominators are the same, each piece is the same size. The fraction with the larger numerator is the larger fraction because numerators tell how many pieces you have.

   b. false; Possible explanation: If the numerators are the same, the number of pieces is the same. As the denominator increases, the size of the pieces decreases, so the fraction with the smaller denominator is the larger fraction.

4. a. Small slices are $\frac{1}{12}$ of a pizza, and large slices are $\frac{1}{8}$ of a pizza; $\frac{3}{12} + \frac{3}{8} = \frac{15}{24} = \frac{5}{8}$ of a pizza.

   b. $1 - \frac{15}{24} = \frac{9}{24}$ of a pizza

   c. $\frac{2}{8} + \frac{9}{12} = 1$; $\frac{4}{8} + \frac{6}{12} = 1$; $\frac{6}{8} + \frac{3}{12} = 1$

5. a. $1\frac{1}{3}$ b. $2\frac{1}{2}$

Answers to Check-Up 2

1. **a.** $\frac{8}{40}$ or $\frac{1}{5}$; 0.2 or 0.20   **b.** $\frac{3}{24}$ or $\frac{1}{8}$; 0.125

2. 0.00006, 0.006, 0.06, 0.60 = 0.6, 6.00

3. Possible answers:

    a. 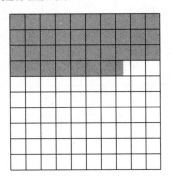   **b.**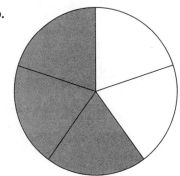

    c.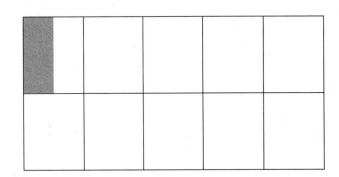

4. **a.** $\frac{375}{1000}$ or $\frac{3}{8}$ **b.** $\frac{6}{10}$ or $\frac{3}{5}$ **c.** $\frac{5}{100}$ or $\frac{1}{20}$

5.
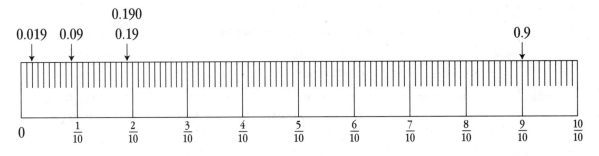

6. **a.**
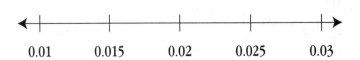

b.

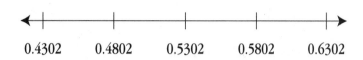

0.3061   0.311   0.3161   0.3211   0.3261

c.

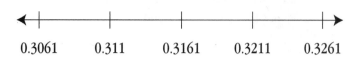

0.4302   0.4802   0.5302   0.5802   0.6302

Answers to Question Bank

1. **a.**

b. 10 is $\frac{10}{100}$ or $\frac{1}{10}$; 25 is $\frac{25}{100}$ or $\frac{1}{4}$; 35 is $\frac{35}{100}$ or $\frac{7}{20}$; 70 is $\frac{70}{100}$ or $\frac{7}{10}$; 85 is $\frac{85}{100}$ or $\frac{17}{20}$

c. 33 or 34 points

d. 60 points

e. 25 points

f. 75 points

g. Taylor should receive a few more points than Miki because $\frac{6}{9}$ of the way up the pole is more than $\frac{5}{8}$ of the way.

2. If you want the most pizza possible, join the first group. In the first group, you would share 6 pizzas among 4 people, so you would receive $1\frac{1}{2}$ pizzas. In the second group, you would share 8 pizzas among 6 people, so you would receive $1\frac{1}{3}$ pizzas.

3. a. $\frac{3}{12}$ or $\frac{1}{4}$

    b. $\frac{5}{20}$ or $\frac{1}{4}$

    c. $\frac{2}{9}$

    d. $\frac{7}{17}$

4. The muffins may not have been the same size to start with.

5. When comparing two decimal numbers that are both less than 1, you need compare place-value amounts. For example, 0.37 is less than 0.6. This is because 0.37 means 37 hundredths or 3 tenths and 7 hundredths, and 0.6 means 6 tenths. Tenths are greater than hundredths, so 0.6 is more than 0.37.

6. The decimal point should not be in these signs. The way the prices are written, bananas cost less than a penny a pound, and the price of the paper is between one and two cents.

7. The Events Night held by Mr. Martinez's and Ms. Swanson's middle-school classes was a success, raising a total of *$645*. The teachers estimated the large turnout of middle-school students included over $\frac{3}{4}$ of the building's student population. Over half of the money, *$330.65*, was earned by the food booths. *215* game tickets were sold, raising *$161.25*, which represented $\frac{1}{4}$ of the money. The tickets were *75* cents each. Most of the money that was raised, *65%*, will go toward paying for the class camping trip, and the other *35%* will be used to pay expenses.

8. a. 6%; In decimal form, 6% is 0.06.

    b. $\frac{1}{25}$; In decimal form, $\frac{1}{25}$ is 0.04; as a percent, it is 4%.

    c. 34%; In decimal form, 34% is 0.34; as a fraction, it is $\frac{34}{100}$ or $\frac{17}{50}$.

9. a. $\frac{1}{4}$ b. $\frac{1}{2}$

Answers to the Unit Test: In-Class Portion

1. a. % shaded: 12%

        % not shaded: 88%

    b. % shaded: 8%

        % not shaded: 92%

    c. % shaded: 30%

        % not shaded: 70%

2. a. $\frac{30}{100}$ or $\frac{3}{10}$; 0.30 or 0.3; 30%

    b. $\frac{55}{100}$ or $\frac{11}{20}$; 0.55; 55%

    c. $\frac{20}{25}$ or $\frac{4}{5}$; 0.80 or 0.8; 80%

    d. $\frac{3}{4}$; 0.75; 75%

    e. $\frac{21}{40}$; 0.525; 52.5% or $52\frac{1}{2}$%

    f. $\frac{5}{30}$ or $\frac{1}{6}$; about 0.167; $16\frac{2}{3}$% or about 16.7%

3. **a.** The shop should charge around $1.25 (1.23125 is $\frac{1}{8}$ of the price of a pizza).

   b. The shop should charge around $0.85 (about 0.82083 is $\frac{1}{12}$ of the price of a pizza).

   c. Answers will vary depending on answers for parts a and b.

4. 120 people; Possible explanations: 20% means 20 out of every 100. You have 6 hundreds, so $6 \times 20 = 120$. Or, 20% can be written as $\frac{1}{5}$, and a fifth of 600 is 120.

5. Answers will vary.

Check-Up 1 is the first assessment piece for this unit. The blackline master for Check-Up 1 is on pages 86 and 87. Below are the scoring rubric one teacher used to assess the check-up and two examples of student work. After each example, the teacher comments about how she assessed the work.

Suggested Scoring Rubric

Question	Possible score	Scoring breakdown
1	5 points	3 points for labeling the marks 2 points for explaining how the labels were determined
2	10 points	2 points for each pair of numbers—1 point for the correct comparison and 1 point for a reasonable explanation of how the numbers were compared
3	3 points	1 point for each correct answer
4	3 points	1 point for recognizing that Julie is correct 2 points for explaining why Julie is correct
5	6 points	2 points for each correct answer—1 point for placing the number 1 in a reasonable location and 1 point for a explaining (in words or with a drawing) how this location was determined
6	5 points	3 points for the correct order 2 points for an explanation in words or with a drawing

Total Possible Score: 32 points

Grading Scale: A: 29–32
                        B: 25–28
                        C: 21–24
                        D: 17–20

Samples 1 and 2

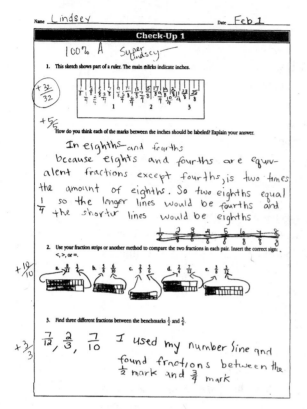

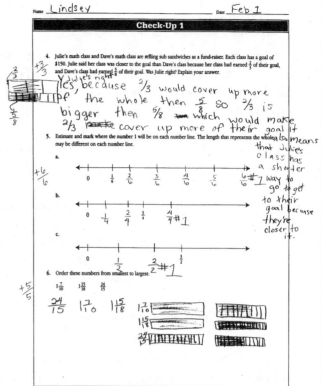

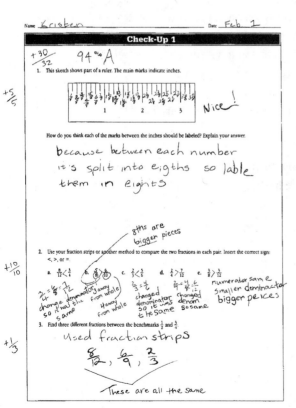

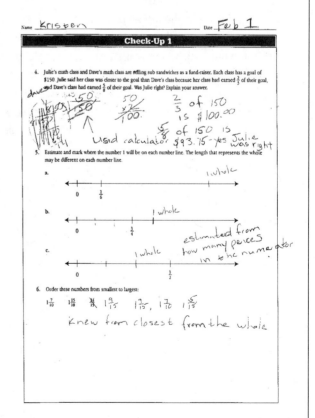

A Teacher's Comments on Sample 1

Lindsey earned all 32 points on her check-up. For question 1, she correctly labeled all of the marks, explained why she labeled the marks as she did, and explained the relationship between fourths and eighths. For question 2, she used the correct symbols to compare the fractions and proved each answer using fraction strips. In questions 3 and 4, she answers the questions correctly and used fraction strips to reason and to justify her answers. Her answers for questions 5 and 6 are also correct. The additional markings she included in questions 5 and her drawings of the strips for question 6 show her reasoning.

The fact that Lindsey used fraction strips as her primary means to solve the problems is fine at this stage in the unit. It is most important that she has a means to solve these types of problems even if her strategy is not what has been traditionally thought of as "sophisticated." As the class continues with this unit, I will continue to encourage Lindsey, and others like her, to think about more sophisticated methods of dealing with fractional relationships.

A Teacher's Comments on Sample 2

Kristen earned 30 of the 32 possible points for her check-up. She earned all of the possible points for questions 1 and 2. Her reasons for question 2 were interesting because she used more than one strategy. For some parts, she found common denominators, and for others, she reasoned about the sizes of the numerators and denominators. She earned only 1 of the three possible points for question 3 because she did not name three *different* fractions. For question 4, I think that Kristen used the strategies we used in class to solve the fund-raiser problems in Investigation 1. Even though she scribbled out some of her work, I can see that she divided 3 into 150 to get 50, and then multiplied 50 by 2 to find that $\frac{2}{3}$ of 150 is 100. Her work for $\frac{5}{8}$ is less clear. She struggled with doing calculations by hand and made some calculation errors. I watched her do the work on her calculator, and then write that $\frac{5}{8}$ would be $93.75, which is correct. Kristen's answers for questions 5 and 6 are correct, and her work makes her statement reasonable.

Kristen incorporated several strategies when reasoning about the problems on the quiz. I need to remind her to be careful and to check over her work before handing it in.

The blackline master for Check-Up 2 is on pages 90–92. Below are the scoring rubric one teacher used to assess the check-up and two examples of student work. After each example, the teacher comments about how she assessed the work.

Suggested Scoring Rubric

Question	Possible score	Scoring breakdown
1	4 points	2 points for each part—1 part for the fraction answer and 1 point for the decimal answer
2	4 points	1 point for stating that 0.6 and 0.60 were equal 3 points for the correct order A point is lost for each number excluded from the list
3	3 points	1 point for each part
4	3 points	1 point for each part
5	5 points	1 point for correctly locating each number
6	3 points	1 point for the correct label in part a Note: I made parts b and c extra credit, awarding a *total* of 1 point for each part. These were very difficult for my students, especially part c.

Total Possible Score: 22 points

Grading Scale: A: 20–22
                   B: 18–19
                   C: 15–17
                   D: 12–14

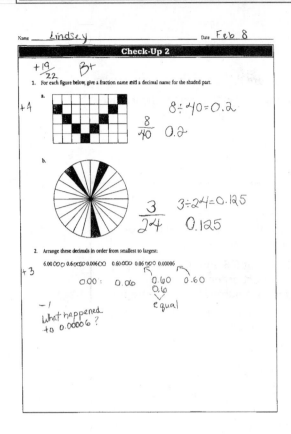

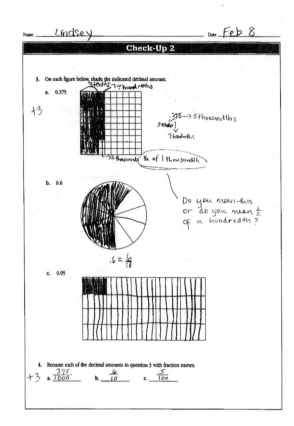

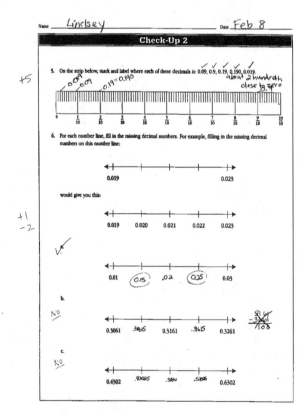

A Teacher's Comments

Lindsey earned 19 of the possible 22 points on this check-up. For question 1, she correctly named the shaded areas with both a fraction name and a decimal name. She lost 1 point for question 2 because she failed to include one of the numbers in her ordered list. For questions 3 and 4, she correctly shades and labels each drawing. She gives hints about her thinking by attempting to explain how she reasoned about part a. She made a small mistake when she wrote that 5 thousandths was $\frac{1}{2}$ of 1 thousandth instead of $\frac{1}{2}$ of 1 hundredth. Because the rest of her work is correct, I did not take off any points for this mistake. She correctly placed each label in question 5, and her added explanations make it clear that she understood the size of the decimals and their locations. She had some trouble with part a of question 6. She had the idea that the marks were 5 units apart but she struggled with the place values of the digits. I gave her 1 out of 3 points for this part; I felt that she could have reasoned through her mistakes, since she was able to label the middle mark correctly. She received no extra credit for parts b and c.

I am a little concerned about Lindsey's struggle with question 6. Because this question was difficult for many of my students, I will take time in class to discuss it. I might make up some questions like this and give them as opening questions for a future lesson.

Name Kristen   Date Feb 8

Check-Up 2

$\frac{+20}{22}$ A   Great!

1. For each figure below, give a fraction name and a decimal name for the shaded part.

+A

a.

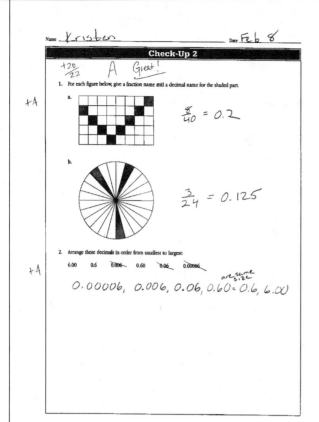

$\frac{8}{40} = 0.2$

b.

$\frac{3}{24} = 0.125$

2. Arrange these decimals in order from smallest to largest:

+A

6.00   0.6   0.006   0.60   0.06   0.00006

$0.00006, \ 0.006, \ 0.06, \ 0.60 = 0.6, \ 6.00$ are same size

---

Name Kristen   Date Feb 8

Check-Up 2

3. On each figure below, shade the indicated decimal amount.

a. 0.375

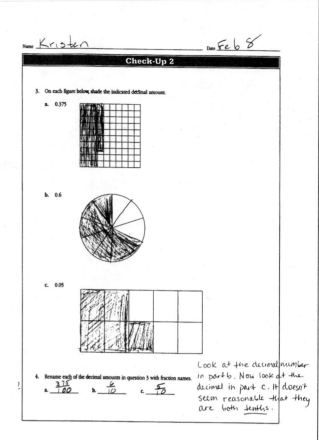

b. 0.6

c. 0.05

4. Rename each of the decimal amounts in question 3 with fraction names.

a. $\frac{375}{100}$ b. $\frac{6}{10}$ c. $\frac{5}{10}$

Look at the decimal number in part b. Now look at the decimal in part c. It doesn't seem reasonable that they are both tenths.

Name Kristen   Date Feb 8

Check-Up 2

5. On the strip below, mark and label where each of these decimals is: 0.09, 0.9, 0.19, 0.190, 0.019.

+5

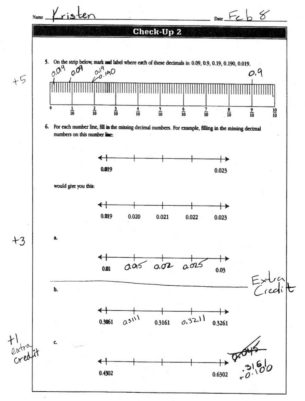

0.09   0.09   0.19   0.190   0.9

0   $\frac{1}{10}$ $\frac{2}{10}$ $\frac{3}{10}$ $\frac{4}{10}$ $\frac{5}{10}$ $\frac{6}{10}$ $\frac{7}{10}$ $\frac{8}{10}$ $\frac{9}{10}$ $\frac{10}{10}$

6. For each number line, fill in the missing decimal numbers. For example, filling in the missing decimal numbers on this number line:

0.019   0.023

would give you this:

0.019   0.020   0.021   0.022   0.023

+3   a.

0.01   0.015   0.02   0.025   0.03

Extra Credit

b.

0.3061   0.3111   0.3161   0.3211   0.3261

+1 extra credit   c.

0.4302   0.6302

.045   .3161   +0.100

A Teacher's Comments

Kristen earned 21 points (including an extra credit point for question 6). She earned all the possible points for questions 1 and 2. For question 2, she earned 2 of the 3 possible points because she shaded five tenths, rather than five hundredths, for part c. She received 2 of the 3 possible points for question 4. Since the fraction she wrote for part c corresponded with her drawing, I considered not subtracting a point. However, I decided to take off a point, because she could have looked at the decimal names for parts b and c and realized that the fractions could not both be tenths. She earned all the possible points for part a of question 6 and received 1 extra credit point for part b.

This check-up shows me that Kristen has made enough sense of the skills in this unit to move on. I would ask her what she was thinking when she shaded part c of question 3, but I am not too concerned about his because of her ACE work and her work in class.

Several teachers have taught the Connected Mathematics program to special education students who are part of an inclusion program. In one school district, special education students study mathematics in a regular classroom with both a mathematics teacher and a special education teacher present. The teachers work together to teach the curriculum. The mathematics teacher leads the lesson, and the special education teacher joins the discourse and works with the students during the Explore phase.

In this school, the teachers have decided to include the special education students in the classroom for everything, including assessment lessons. For formal assessment, the teachers work together to create special assessment pieces that are similar to the assessments the other students are doing. The teachers' goals are to offer challenging, yet realistic, problems that allow them to assess what their students know and what they are struggling with. Below is an example of a special education student's work on an altered check-up. After the sample, we include the teachers' comments about why they changed each question and how they assessed the student's work.

Sample

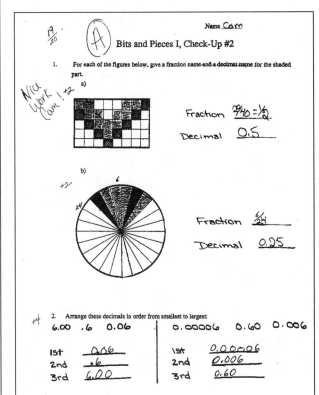

$\frac{19}{20}$

Name **Cam**

(A) Bits and Pieces I, Check-Up #2

1. For each of the figures below, give a fraction name and a decimal name for the shaded part.

Nice Work Cam! +2

a)

Fraction $\frac{20}{40} = \frac{1}{2}$

Decimal **0.5**

+2 b)

6

24

Fraction $\frac{6}{24}$

Decimal **0.25**

+4 2. Arrange these decimals in order from smallest to largest:

6.00   .6   0.06 | 0.00006   0.60   0.006

1st **0.06**   1st **0.00006**
2nd **.6**   2nd **0.006**
3rd **6.00**   3rd **0.60**

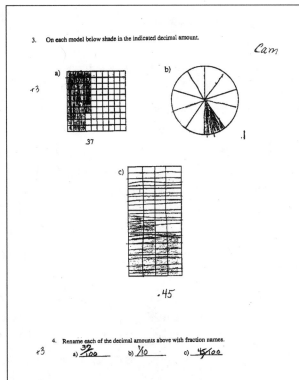

3. On each model below shade in the indicated decimal amount.

Cam

a)
+3
.37

b)
.1

c)
.45

4. Rename each of the decimal amounts above with fraction names.
+3 a) $\frac{37}{100}$ b) $\frac{1}{10}$ c) $\frac{45}{100}$

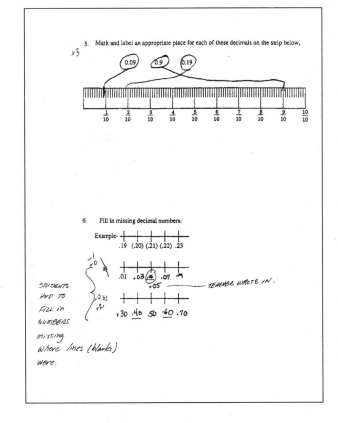

5. Mark and label an appropriate place for each of these decimals on the strip below,
×3
0.09 0.9 0.19

| 1/10 | 2/10 | 3/10 | 4/10 | 5/10 | 6/10 | 7/10 | 8/10 | 9/10 | 10/10 |

6. Fill in missing decimal numbers:

Example- .19 (.20) (.21) (.22) .23

a) .01 .03 .04 .07
         .05   TEACHER WROTE IN.

STUDENTS
HAD TO
FILL IN
NUMBERS
missing
where lines (blanks)
were.

b) .30 .40 .50 .60 .70

The Teachers' Comments

Question 1

We altered the drawings to represent friendlier fractions ($\frac{1}{2}$ instead of $\frac{1}{5}$ and $\frac{1}{4}$ instead of $\frac{1}{8}$). We also provided labeled answer blanks to remind students that there are two parts to the question.

Cam answered both questions correctly. He demonstrated that he knows that $\frac{20}{40}$ is equal to $\frac{1}{2}$, but we wonder if he also knows that $\frac{6}{24}$ is equal to $\frac{1}{4}$. One of us will ask him when we hand back the papers.

Question 2

We simplified this question by asking students to work with three numbers at a time, rather than six. The sheer number of items in a question can be so overwhelming for special needs students that they cannot even begin the problem.

Cam answered this question correctly. If we ask a similar question in the future, we might increase the number of items to four.

Questions 3 and 4

From the work in class, we knew that our special education students would not be able to deal with the thousandths in part a (0.375). By changing the question, we could see whether students could make sense of hundredths in both situations where a grid is given (part a) and in situations where it is not (part c).

Cam answered these questions correctly. We found it interesting that, in part c, he divided the rectangle into 100 pieces and then shaded.

Question 5

Once again, we decreased the number of items. We deleted 0.019 because of the difficulty students have in dealing with thousandths. We deleted 0.190, one of the two equivalent decimals, because dealing with equivalent decimals in situations like this has proven very confusing for students. We believe that we need to help our special education students make sense of this idea but do not feel that an assessment situation is the best way to deal with this problem.

Question 6

As it was originally written, this question is very difficult for students. We altered the problem by using only tenths and hundredths, numbers we knew our special needs students could handle. We also simplified the problem by writing in some of the numbers.

Cam struggled with part a, in which the marks increased by 0.02. Many of the other special education students struggled with both part a and part b.

Overall, Cam did well on the check-up. Not all of the special education students were as successful as he. We included his work to show how well special education students can do when they are given some scaffolding. Each time we alter an assessment piece, we must consider whether we are asking enough of our special students or whether we should expect more. For Cam and others, the success they are having gives them encouragement to try harder. Many of these students have had little previous success learning mathematics because of the focus on memorization. With a focus on understanding and making sense of ideas, these students are realizing that they can learn mathematics.

Blackline Masters

Fraction Strips

halves	

thirds	

fourths	

fifths	

sixths	

eighths	

ninths	

tenths	

twelfths	

Brownie Pans

Justin's Garden

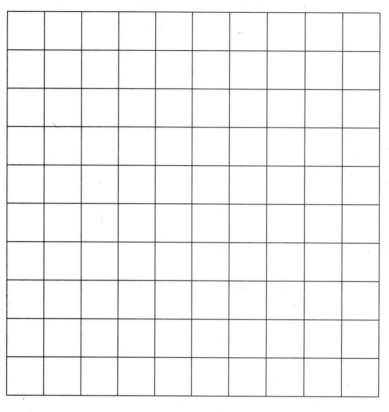

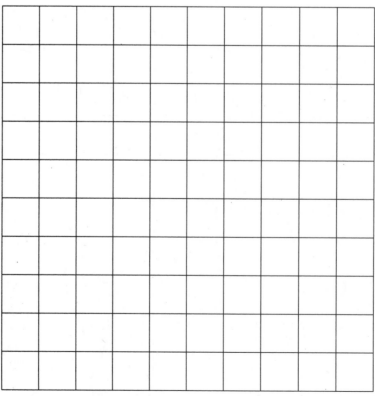

Hundredths Grid

ACE Question 9

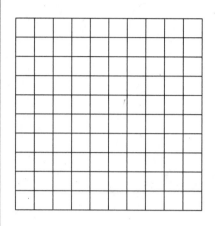
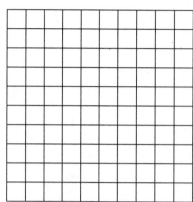
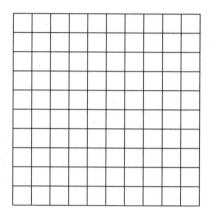

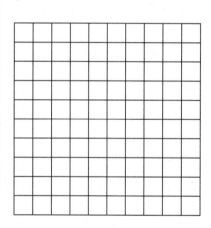
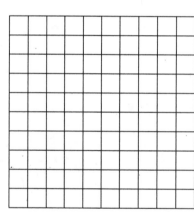
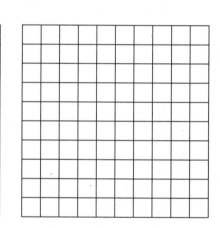

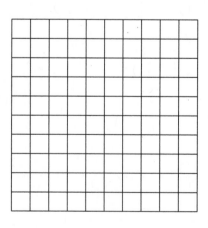
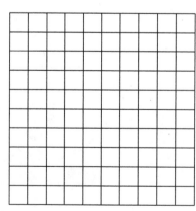
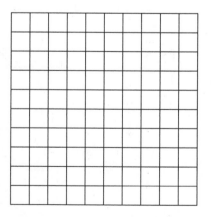

Twelve Hundredths Grids

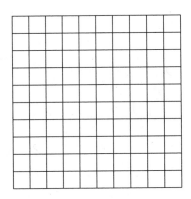 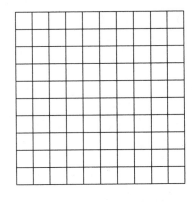 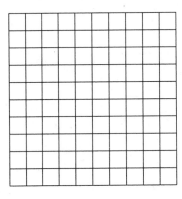

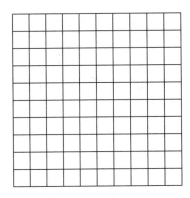 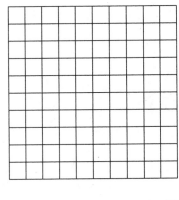 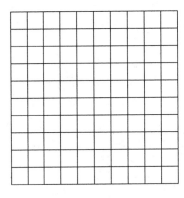

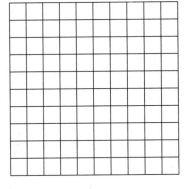 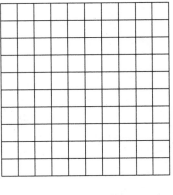 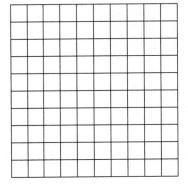

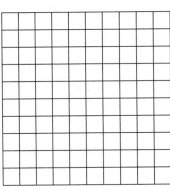 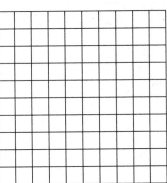 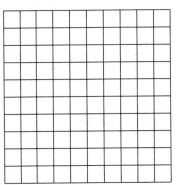

Fraction Strips with Hundredths Strip

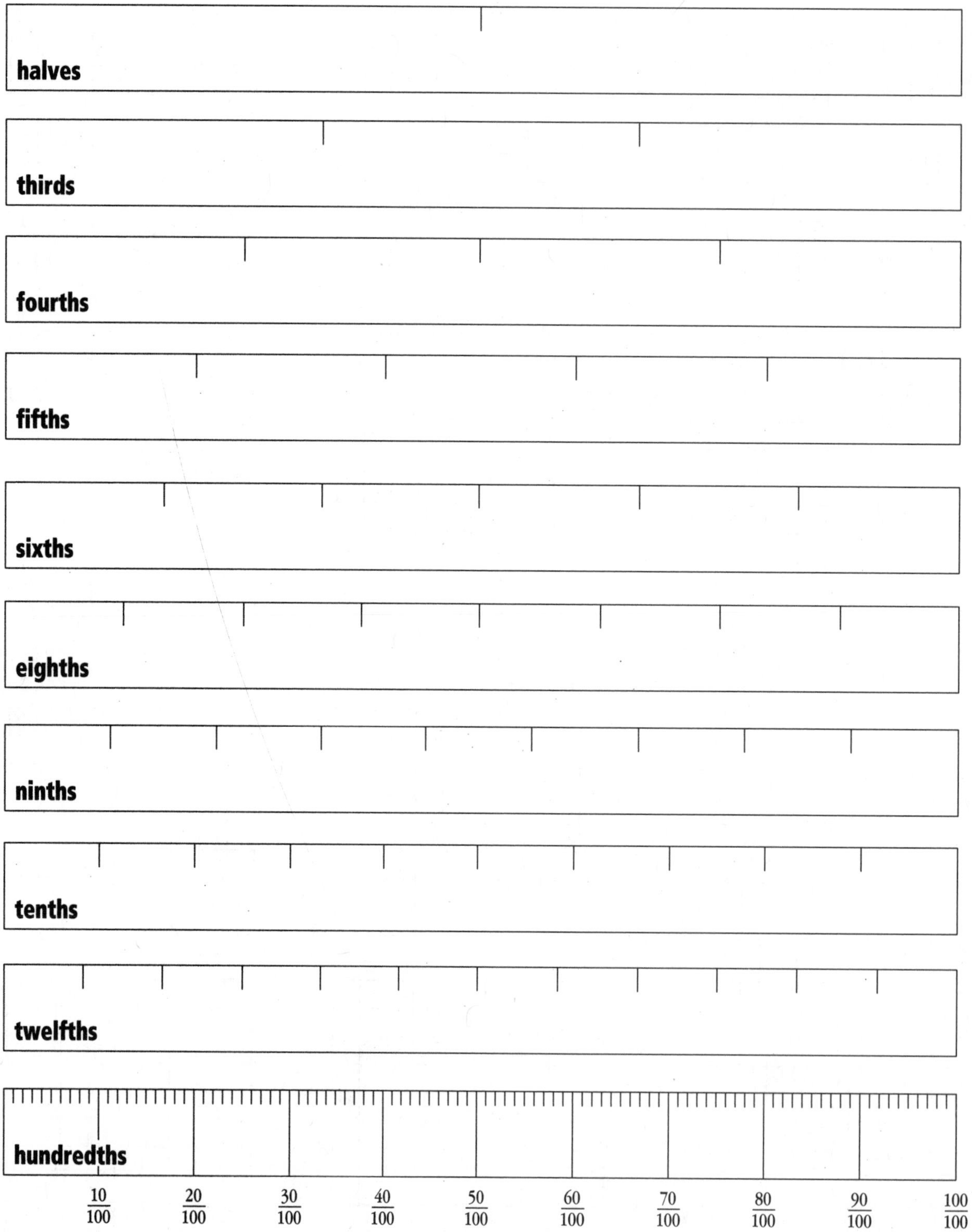

halves

thirds

fourths

fifths

sixths

eighths

ninths

tenths

twelfths

hundredths

$\frac{10}{100}$ $\frac{20}{100}$ $\frac{30}{100}$ $\frac{40}{100}$ $\frac{50}{100}$ $\frac{60}{100}$ $\frac{70}{100}$ $\frac{80}{100}$ $\frac{90}{100}$ $\frac{100}{100}$

100 Cats

Alex	Boots	Diva	Fuzzy	Hanna	Matilda	Newton	Ravena	Smokey	Thomas
Amanda	Bosley	Duffy	Gabriel	Harmony	Melissa	Peanut	Reebo	Smudge	Tiger
Augustus	Bradley	Ebony Kahlua	George	Jinglebob	Mercedes	Peebles	Samantha	Snowy	Tigger
Baguera	Buffy	Elizabeth	Ginger	Kali	Midnight	Pepper	Sassy	Sparky	Ting
Black Foot	Charcoal	Emma	Gizmo	Kiki	Millie	Pink Lady	Scooter	Speedy	Tom
Blacky	Chelsea	Emmie	Gracie	Kitty	Miss Muppet	Pip	Sebastian	Stinky	Tomadachi
Blue	Chessis	Ethel	Gray Kitty	Lady	Mittens	Precious	Seymour	Sweet Pea	Treasure
Bob	Chubbs	Feather	Grey Boy	Libby	Molly	Priscilla	Shiver	Tabby	Wally
Boggie	Cookie	Fire Smoke	Grey Girl	Lucky	Momma Kat	Prissy	Simon	Tabby Burton	Weary
Boo	Dana	Fluffy	Grey Poupon	Lucy	Nancy Blue	Ralph	Skeeter	Terra	Ziggy

Pet Ransom			
	Percent	Decimal	Fraction
$2000 and up	18%		
From $1500 to $1999		0.03	
From $1000 to $1499			$\frac{3}{100}$
From $500 to $999	25%		
From $1 to $499		0.31	
Nothing			$\frac{1}{5}$

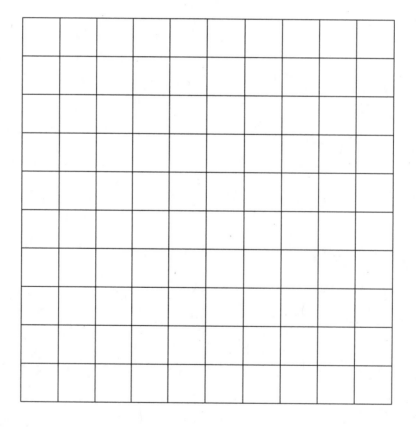

ACE Questions 17–19

17.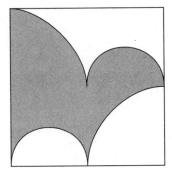

18.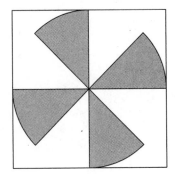

19.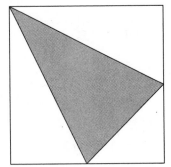

Write a short—but clever and informative—announcement to report the progress of the sixth-grade poster sale after two days. Be sure to mention what part of the sales goal of $300 had been reached and what part remained to be raised.

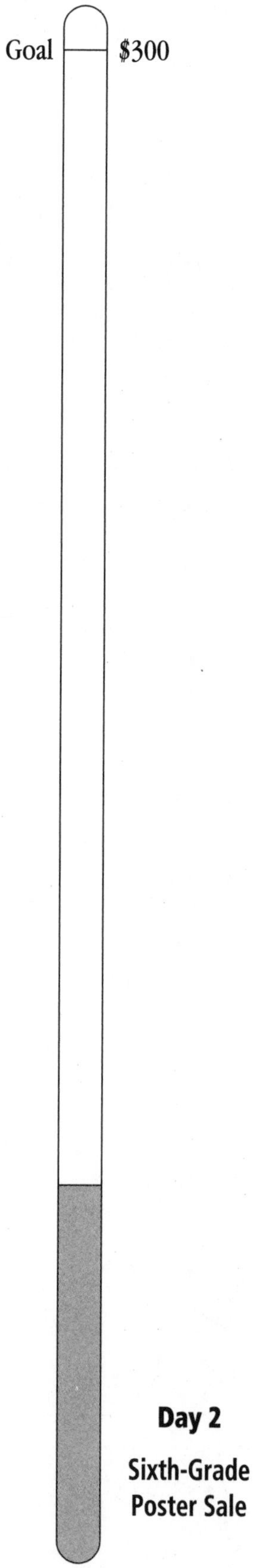

Goal — $300

Day 2

Sixth-Grade
Poster Sale

Goal — $300   Goal — $300   Goal — $300   Goal — $300   Goal — $300

Day 2   **Day 4**   **Day 6**   **Day 8**   **Day 10**

Start with nine $8\frac{1}{2}$-inch strips. Fold the strips to show halves, thirds, fourths, fifths, sixths, eighths, ninths, tenths, and twelfths. Mark the folds in the strips with a pencil so you can see them more easily.

Use your strips to estimate the sixth-grade class's progress after two, four, six, eight, and ten days.

Thermometers for All Grades

Transparency 1.3A

Goal
$300
Day 10
Sixth-Grade
Poster Sale
Goal
$400
Day 10
Seventh-Grade
Popcorn Sale
Goal
$240
Day 10
Eighth-Grade
Calendar Sale

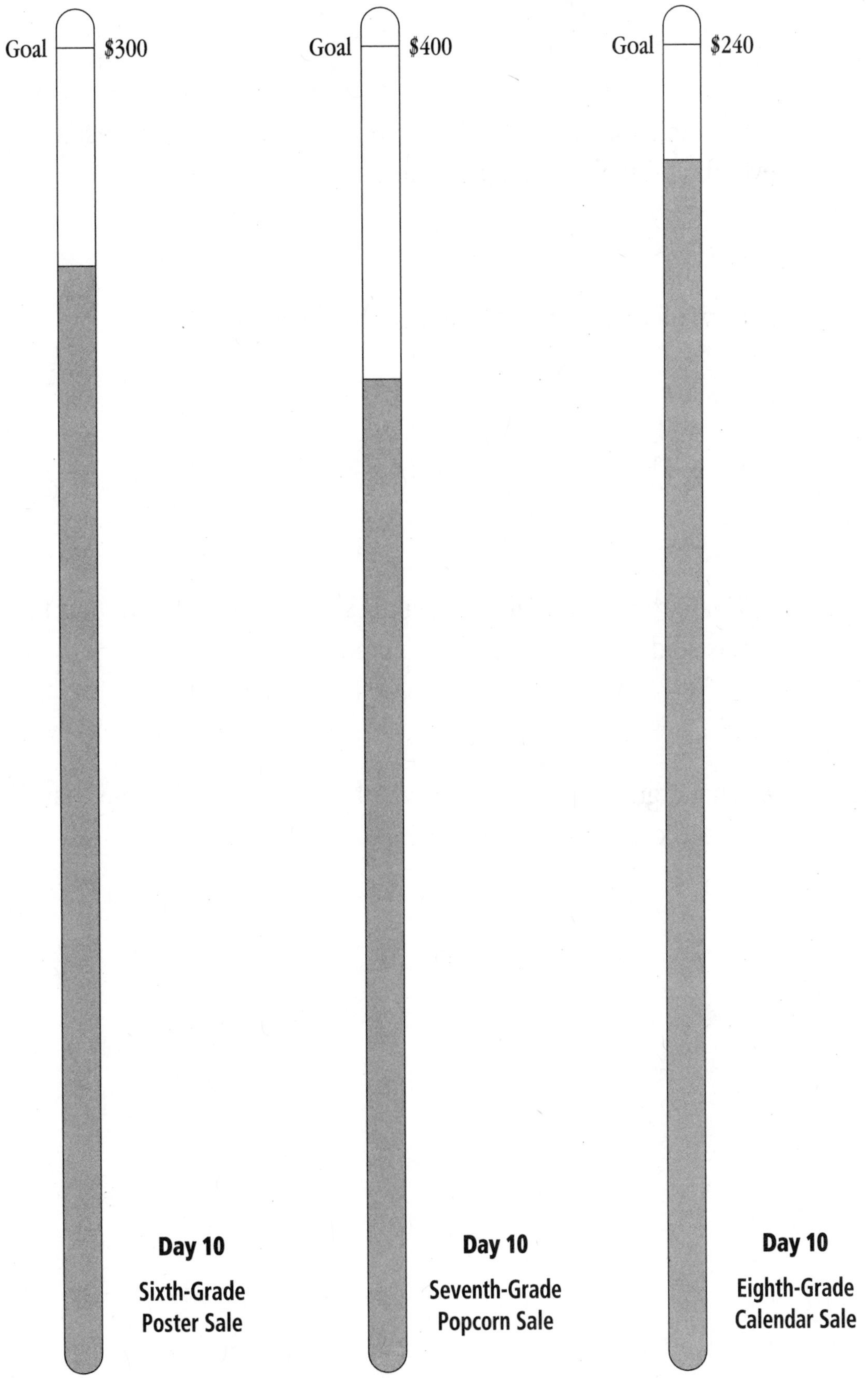

Use the fraction strips you made in Problem 1.2 to investigate the seventh and eighth graders' claims.

A. How much money did each grade raise?

B. What fraction of the goal did each grade reach?

C. What argument could the eighth graders use to claim that their class did better than the sixth grade?

D. What argument could the seventh graders use to claim that their class did better than the sixth grade?

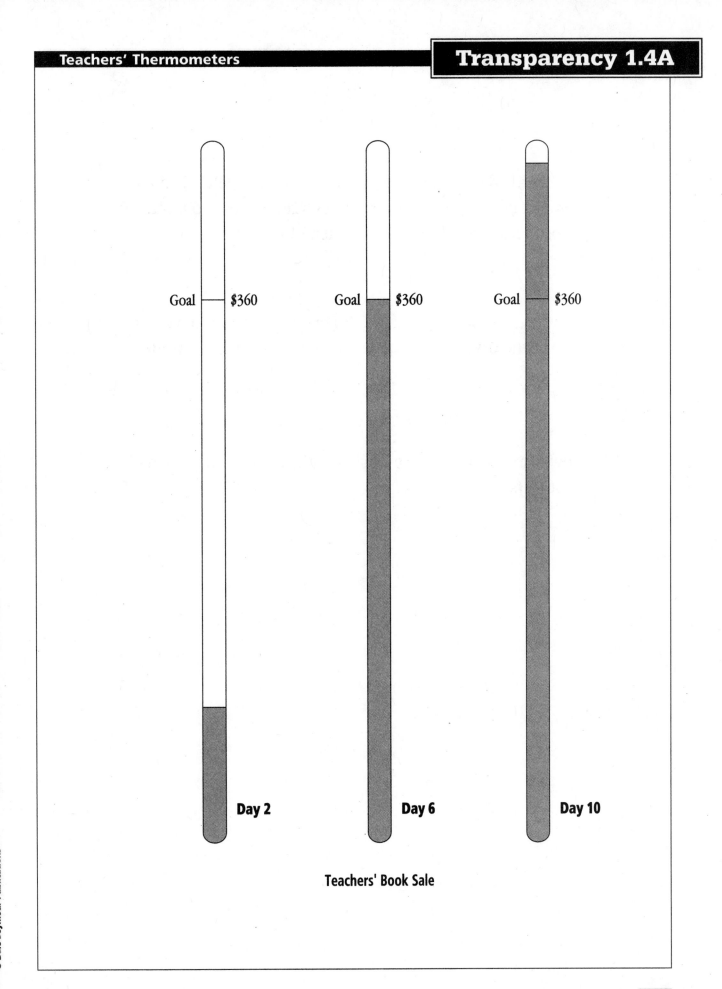

Goal — $360   Goal — $360   Goal — $360

Day 2   **Day 6**   **Day 10**

Teachers' Book Sale

A. Notice that the teachers used a shorter thermometer than the students did to report their progress. Can you use your fraction strips to measure these thermometers? Explain.

B. What fraction of their goal did the teachers reach at the end of each of the days shown? Explain how you determined your answers.

C. How many dollars did the teachers raise by the end of each of these days?

Problem 1.5

Transparency 1.5

Each strip below is divided into a different number of equal-length parts. On your copy of Labsheet 1.5, label each of the marks on the strips with fraction names in symbolic form. The label for a mark should represent the fraction of the strip to the left of the mark.

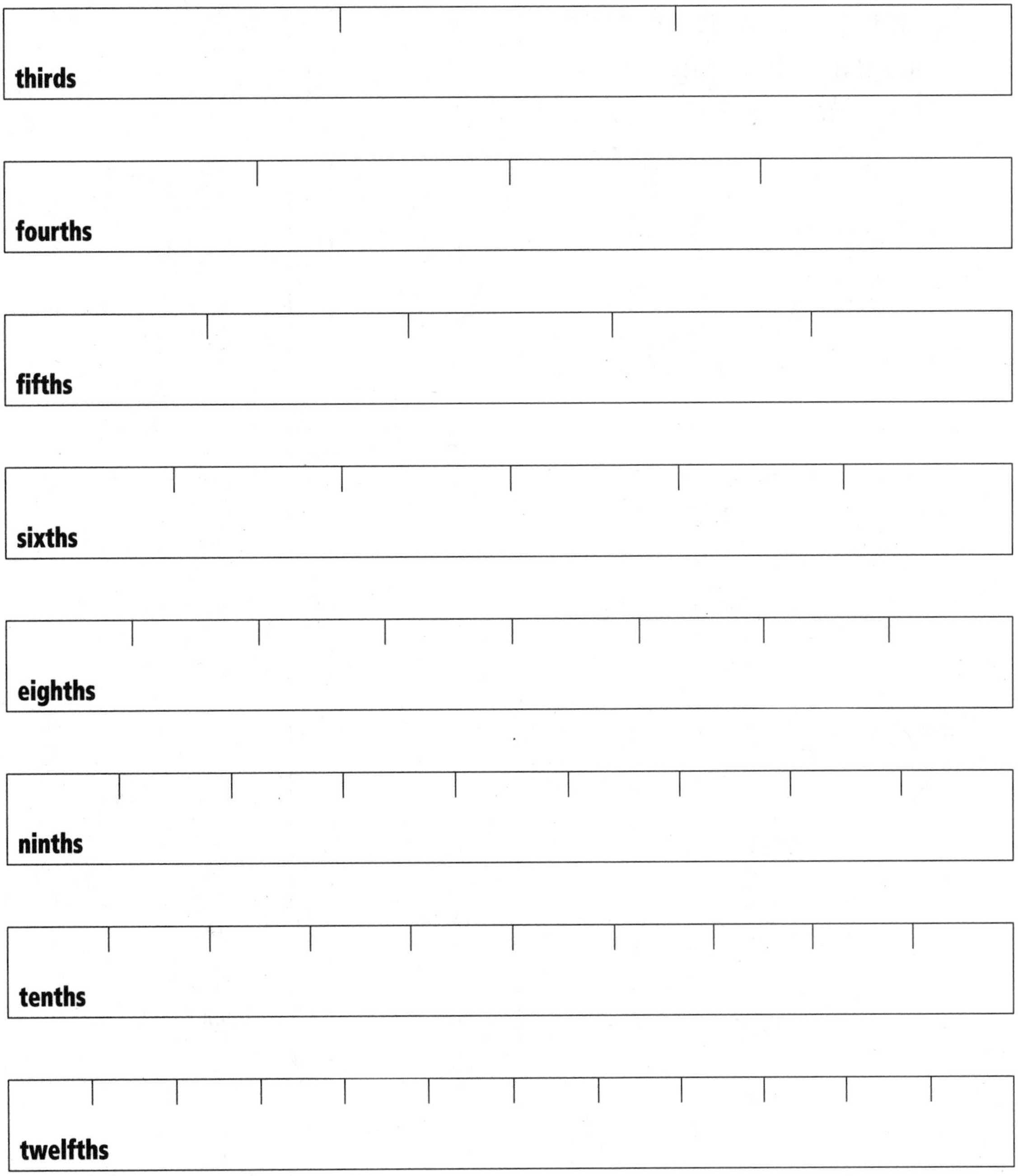

thirds
fourths
fifths
sixths
eighths
ninths
tenths
twelfths

A. Which of the three teachers do you agree with? Why?

B. How could the teacher you agreed with in part A prove his or her case?

Goal — $360

Day 4

Teachers'
Book Sale

The fraction strips on the left below show $\frac{2}{3}$ and three fractions equivalent to $\frac{2}{3}$. The strips on the right show $\frac{3}{4}$ and three fractions equivalent to $\frac{3}{4}$. Look for patterns that will help you find other equivalent fractions.

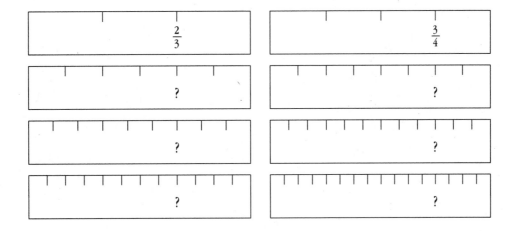

A. What are the three fractions shown that are equivalent to $\frac{2}{3}$? Name three more fractions that are equivalent to $\frac{2}{3}$.

B. What are the three fractions shown that are equivalent to $\frac{3}{4}$? Name three more fractions that are equivalent to $\frac{3}{4}$.

C. What pattern do you see that can help you find equivalent fractions?

$$\frac{1}{3} = \frac{5}{15}$$

A. Make a number line as illustrated below. When you find another name for a mark you have already labeled, record the new name below the first name.

B. Look for patterns in your finished number line. Record your findings.

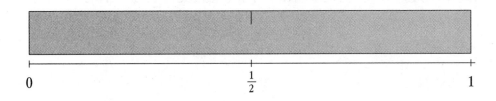

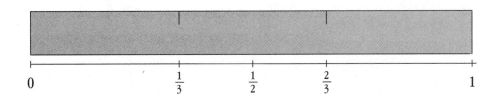

A. Decide whether each fraction below is between 0 and $\frac{1}{2}$ or between $\frac{1}{2}$ and 1.

$$\frac{1}{5} \quad \frac{2}{3} \quad \frac{8}{10} \quad \frac{3}{12} \quad \frac{3}{5} \quad \frac{5}{6} \quad \frac{5}{8} \quad \frac{4}{5} \quad \frac{3}{8} \quad \frac{3}{4} \quad \frac{2}{9} \quad \frac{7}{12} \quad \frac{1}{3}$$

B. Decide whether each fraction from part A is closest to 0, $\frac{1}{2}$, or 1. Record your information in a table.

C. Explain your strategies for comparing fractions to 0, $\frac{1}{2}$, and 1.

D. Use benchmarks and other strategies to help you write the fractions from part A in order from smallest to largest.

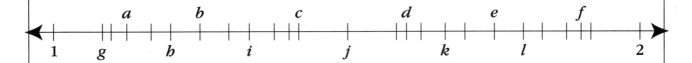

A. Use the fraction strips from Labsheet 1.5 to find as many labels as you can for each of the lettered points. For each point, record the letter and the fraction labels.

B. Copy the number line onto a sheet of paper. Mark and label a point fitting each description below. Do not use points that are already marked.

1. a point close to, but larger than, 1

2. a point close to, but smaller than, $1\frac{1}{2}$

3. a point close to, but larger than, $1\frac{1}{2}$

4. a point close to, but smaller than, 2

Use the squares on Labsheet 3.1 as models for pans of brownies.
Show the cuts you would make to divide a pan of brownies into

A. 15 equal-size large brownies

B. 20 equal-size medium brownies

C. 30 equal-size small brownies

A. Do you think Samantha, Romero, and Harold should make small, medium, or large brownies?

B. If they make brownies of the size you chose in part A, how much of each ingredient will they need to make enough to serve a brownie to each person at camp?

C. Describe the strategy you used to get your answer to part B.

Chunky Brownies with a Crust

$1\frac{1}{4}$ cups flour

$\frac{1}{4}$ cup sugar

$\frac{1}{2}$ cup cold butter or margarine

1 14-ounce can sweetened condensed milk

$\frac{1}{4}$ cup unsweetened cocoa

1 egg

1 teaspoon vanilla

$\frac{1}{2}$ teaspoon baking powder

1 7-ounce bar milk chocolate, broken into small chunks

$\frac{3}{4}$ cup chopped nuts (optional)

Preheat the oven to 350 degrees. In medium bowl, combine 1 cup of flour and the sugar. Cut in the margarine or butter until crumbly. Press the mixture firmly into the bottom of a 10-by-10-inch baking pan. Bake 15 minutes. Meanwhile, in a large mixing bowl, beat the sweetened condensed milk, the cocoa, the egg, the remaining flour, the vanilla, and the baking powder. Stir in the nuts and chocolate chunks. Spread over the prepared crust. Bake 20 minutes or until the center is set. Cool. Sprinkle with confectioner's sugar, if desired. Store tightly covered at room temperature. Makes 15 large, 20 medium, or 30 small brownies.

Brownie Table			
	Small (30)	Medium (20)	Large (15)
Batches for 240	8 batches	12 batches	16 batches
Cups flour	10	15	20
Cups sugar	2	3	4
Cups butter	4	6	8
Cans milk	8	12	16
Cups cocoa	2	3	4
Eggs	8	12	16
Tsp. vanilla	8	12	16
Tsp. baking powder	4	6	8
No. 7-oz choc. bars	8	12	16
Cups nuts	6	9	12

Here are the family's requirements for the garden.

- Justin's father wants to be sure potatoes, beans, corn, and tomatoes are planted. He wants twice as much of the garden to be planted in corn as potatoes. He wants three times as much land planted in potatoes as tomatoes.

- Justin's sister wants cucumbers in the garden.

- Justin's brother wants carrots in the garden.

- Justin's mother wants eggplant in the garden.

- Justin wants radishes in the garden.

Use Labsheet 4.1 to make a suitable plan for the garden. Write a description of the garden you plan. Name the fraction of the garden space that will be allotted to each kind of vegetable as part of your description. Explain how your garden will satisfy each member of Justin's family.

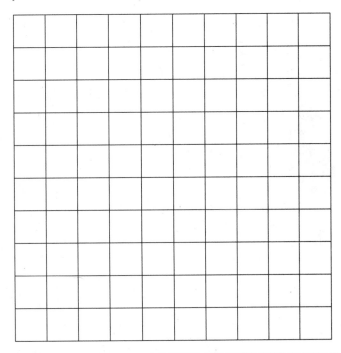

Look back at the original plan you drew for Justin's garden. Write each of the fractional parts for the vegetables in your plan as a decimal.

Fraction names for place-value groups	Decimal names for place-value groups		
Ten thousands	10,000	10,000	
Thousands	1000	1000	
Hundreds	100	100	
Tens	10	10	
Ones	1	1	
Tenths	$\frac{1}{10}$	0.1	
Hundredths	$\frac{1}{100}$	0.01	
Thousandths	$\frac{1}{1000}$	0.001	
Ten thousandths	$\frac{1}{10,000}$	0.0001	

A. Rename each of these fraction benchmarks as a decimal.

1. 0   **2.** $\frac{1}{4}$ **3.** $\frac{1}{2}$ **4.** $\frac{3}{4}$ **5.** 1

B. Now use the decimal benchmarks and other strategies that make sense to you to help you to order each set of numbers from smallest to largest.

1. 0.23   0.28   0.25

2. 2.054   20.54   2.54

3. 0.78   0.708   0.078

C. For each of the three decimals in parts 1 and 3 of question B, give the name of the decimal in words, and tell which benchmark the number is nearest. Give the benchmark as a fraction and as a decimal. Organize your work in a table like the one shown below. For each decimal, explain your reasoning.

Number	Name in words	Nearest decimal benchmark	Nearest fraction benchmark	Reasoning
0.23				
0.28				
0.25				
0.78				
0.708				
0.078				

Play the Distinguishing Digits puzzles with your group. Record the strategies you use to solve the puzzles.

Mystery Number

1 __ . __ __ __

Clue 1 The digit in the thousandths place is double the digit in the ones place.

Clue 2 The digit in the tenths place is odd, and it represents the sum of the digits in the tens place and the thousandths place.

Clue 3 There are exactly two odd digits in the Mystery Number.

Clue 4 The digit in the hundredths place is three times the digit in the ones place.

The coach has three players to choose from to shoot the free throw. In their pregame warm-ups:

• Angela made 17 out of 25 free throws

• Emily made 15 out of 20 free throws

• Carma made 7 out of 10 free throws

Which player should the Portland coach select to shoot the free throw? Explain your reasoning.

Work with your group to find a way to use the fraction strips to help you estimate each of the fractions represented on the halves, thirds, fourths, fifths, sixths, eighths, ninths, tenths, and twelfths fraction strips as decimals.

You might think about doing this by comparing the marks on each fraction strip to marks on the hundredths strip.

For example, to find a decimal name for $\frac{5}{12}$, you can find the mark on the hundredths strip that is nearest to the length $\frac{5}{12}$, since hundredths can easily be written as decimals. Since the mark at $\frac{42}{100}$ on the hundredths strip is the closest mark to $\frac{5}{12}$ on the twelfths strip, $\frac{5}{12}$ is approximately equal to 0.42.

Sometimes it is easier to look at the tenths strip. For example, $\frac{1}{2}$ on the fraction strip is at the same mark as $\frac{5}{10}$ on the tenths strip, so $\frac{1}{2}$ is equivalent to 0.5.

On Labsheet 5.2, label each mark on the halves, thirds, fourths, fifths, sixths, eighths, ninths, tenths, and twelfths fraction strips with an approximate decimal representation. Be prepared to explain your answers.

The students had 24 boxes for packing the food they collected. They wanted to share the supplies equally among the families who would receive the boxes. They had small bags and plastic containers to use to repack items for the individual boxes.

The students collected the following items:

48 tins of cocoa mix	6 pounds of Swiss cheese
72 boxes of powdered milk	3 pounds of hot pepper cheese
264 boxes of juice	7 pounds of peanuts
120 boxes of granola bars	5 pounds of popcorn kernels
36 pounds of wheat crackers	475 apples
18 pounds of peanut butter	195 oranges
12 pounds of cheddar cheese	

A. How much of each item should the students include in each box? Explain your reasoning.

B. What operation (+, −, ×, ÷) did you use to find your answers? Why did this operation work?

C. How can your calculator help you decide how to distribute the food items?

Using the database and Labsheet 6.1, mark all the cats that are female on one chart and all the cats that are kittens on another chart. When you are finished, answer the following questions.

A. What fraction of the cats are female? Write the fraction as a decimal and a percent.

B. What fraction of the cats are male? Write the fraction as a decimal and a percent.

C. What do you notice about the combined percentage of female and male cats?

D. What fraction of the cats are kittens? Write the fraction as a decimal and a percent.

E. What fraction of the cats are adults? Write the fraction as a decimal and a percent.

F. What do you notice about the combined percentage of kittens and adult cats?

Cat	Gender	Age (yrs)	Weight (lbs)	Eye color	Pad color
Alex	m	18	11	green	black
Amanda	f	4.5	9.75	blue	gray
Augustus	m	2	10	yellow/green/blue	pink/black
Baguera	m	0.17	13	yellow	brown
Black Foot	m	0.33	1.5	yellow	gray
Blacky	f	1	5	yellow	gray/black
Blue	f	0.25	2	green	gray
Bob	f	4	12	green	black
Boggie	m	3	10	green	pink
Boo	m	3.5	10.75	yellow/green	brown
Boots	m	0.25	3	brown	black
Bosley	m	0.33	1.5	yellow/brown	pink
Bradley	m	0.6	11	yellow	pink/gray
Buffy	m	0.75	8	blue/green	pink
Charcoal	m	11	12	yellow	black
Chelsea	f	2	9	yellow	black
Chessis	f	1.5	6	green	brown
Chubbs	m	1	7	green	pink
Cookie	f	4	9	gold	black
Dana	f	10	8	green	black
Diva	f	3.5	11	green	pink
Duffy	m	1	9	yellow/green	black
Ebony Kahlua	m	1.5	15	blue	brown
Elizabeth	f	10	9	green	pink
Emma	f	4	9.25	gold	pink
Emmie	f	4	7	green	black
Ethel	f	5	8	green	black
Feather	m	2.5	13	green	pink
Fire Smoke	f	0.25	2.5	green/brown	pink
Fluffy	f	5	10	green	pink
Fuzzy	f	1.25	2	green	pink
Gabriel	m	1	7	blue	white
George	m	12	14.5	green	black
Ginger	f	0.2	2	yellow/green	pink
Gizmo	m	4	10	yellow	black
Gracie	f	8	12	green	pink
Gray Kitty	f	3	9	green	gray

You may want to use fraction strips or hundredths squares to help you to think about these questions.

A. 1. Rewrite the text on the sign for leashes so that the discount is shown as a *fraction off* the original price of the leashes.

    2. What will a $10.00 leash cost after the discount?

B. 1. Rewrite the text on the sign for pet carriers so that the discount is shown as a *percent off* the original price of the carriers.

    2. What will a $10.00 pet carrier cost after the discount?

C. 1. Rewrite the text on the pet food sign so that the discount is shown as a *percent off* the original price of pet food.

    2. Now, write the discount as a *percent* of the original price customers will pay.

    3. Rewrite the discount as a *fraction* of the original price customers will pay.

    4. Rewrite the discount as a *fraction off* the original price customers will pay.

    5. What will $10.00 worth of pet food cost after the discount?

D. 1. Rewrite the text on the pet treats sign so that the discount is shown as a *decimal.*

    2. What will $10.00 worth of pet treats cost after the discount?

Labsheet 6.3 contains the table below and a hundredths grid.

A. Fill in the missing information in your table.

B. Shade in the hundredths grid with different colors or shading styles to show the percent responding to each of the six choices. Add a key to your grid to show what each color or type of shading represents. When you finish, the grid should be completely shaded. Explain why.

	Percent	Decimal	Fraction
$2000 and up	18%		
From $1500 to $1999		0.03	
From $1000 to $1499			$\frac{3}{100}$
From $500 to $999	25%		
From $1 to $499		0.31	
Nothing			$\frac{1}{5}$

Preference	Out of 100 dog owners	Out of 100 cat owners
Human food only	50	27
Pet food only	30	58
Human and pet food	20	15

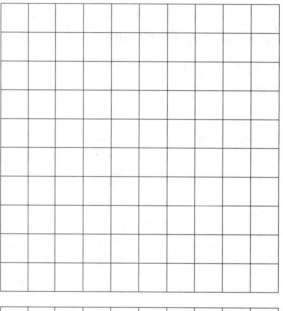

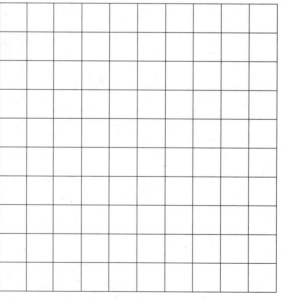

Consider the results of the survey.

Preference	Out of 150 dog owners	Out of 200 cat owners
Human food only	75	36
Pet food only	45	116
Human and pet food	30	48

A. What kind of food is favored by the greatest number of dogs, according to their owners? Write this number as a fraction, a decimal, and a percent of the 150 dog owners surveyed.

B. What choice is favored by the greatest number of cats, according to their owners? Write this number as a fraction, a decimal, and a percent of the 200 cat owners surveyed.

C. What percent of dog owners reported that their dogs liked either human food only or pet food only? Write this percent as a fraction and a decimal.

D. What percent of cat owners reported that their cats liked either human food only or pet food only? Write this percent as a decimal and a fraction.

Dear Family,

The next unit in your child's course of study in mathematics class this year is *Bits and Pieces I*. It is the first of two units to develop understanding of fractions, decimals, and percents—the ideas that are at the heart of the middle-grades experience with number concepts.

Bits and Pieces I focuses on developing a deep understanding of rational numbers, rather than on rules and formulas for computation. Computations with fractions, decimals, and percents will be the focus of the second fraction unit, *Bits and Pieces II*. In this unit, your child will learn the meanings of fractions, decimals, and percents, and will become comfortable moving among these three representations of rational numbers. Your child will work on problems that reflect real-world situations and that involve writing, comparing, and ordering fractions and decimals.

This unit makes use of concrete models, such as fraction strips, number lines, and grids to help students reason about fractions. Skill with estimating and comparing fractions is developed through a set of benchmark fractions and their decimal and percent equivalents. These benchmark fractions are those that occur often in real-world situations.

Fraction	$\frac{1}{10}$	$\frac{1}{8}$	$\frac{1}{5}$	$\frac{1}{4}$	$\frac{1}{3}$	$\frac{1}{2}$	$\frac{2}{3}$	$\frac{3}{4}$
Decimal	0.10	0.125	0.2	0.25	0.33	0.5	0.67	0.75
Percent	10%	12.5%	20%	25%	33%	50%	$66\frac{2}{3}$%	75%

It is important that you do not show your child rules or formulas for working with fractions. This unit helps students to discover these rules for themselves and to develop a firm understanding of why these rules work. You can help your child with his or her work for this unit in several ways:

- Talk to your child about the ways you use fractions, decimals, and percents.
- Point out examples of how fractions, decimals, and percents are used in newspapers, magazines, radio, and television.
- Look over your child's homework and make sure all questions are answered and that explanations are clear.

As always, if you have any questions or concerns about this unit or your child's progress in class, please feel free to call. All of us here are interested in your child and want to be sure that this year's mathematics experiences are enjoyable and promote a firm understanding of mathematics.

Sincerely,

Distinguishing Digits Cards

Mystery Number

_ _ _

Problem 1

Clue

The number has no repeated digits.

Problem 2

Clue

The number has no repeated digits.

Problem 1

Clue

All of the digits in the number are odd.

Problem 2

Clue

All of the digits in the number are even.

Problem 1

Clue

The digit in the ones place is greater than the digit in the tens place.

Problem 2

Clue

The digit in the hundreds place is 2 times the digit in the tens place.

Problem 1

Clue

The digit in the ones place is less than the digit in the hundreds place.

Problem 2

Clue

The digit in the tens place is 2 times the digit in the ones place.

Problem 1

Clue

The sum of the digits is 9.

Problem 2

Mystery Number

_ _ _

Problem 2

Mystery Number

_ _ _ , _ _ _

Problem 3

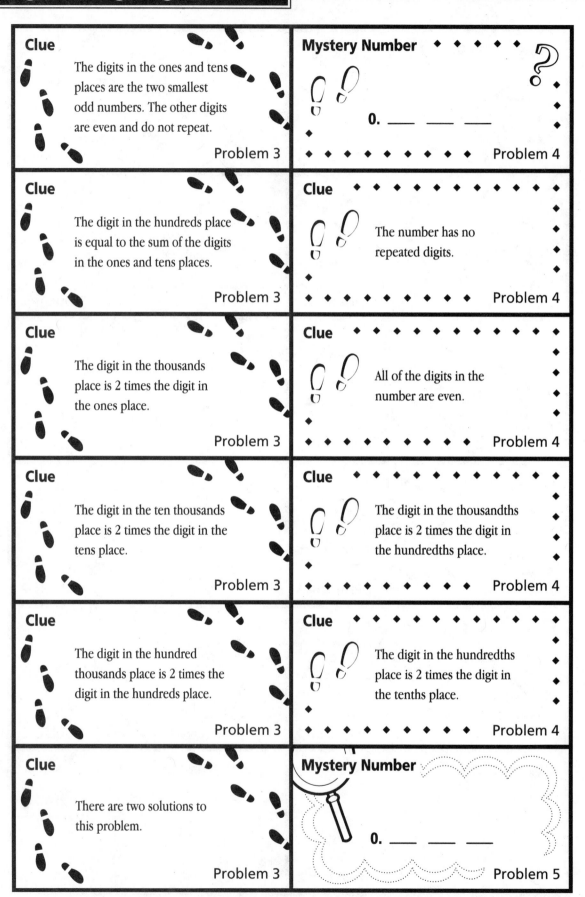

Clue

The digits in the ones and tens places are the two smallest odd numbers. The other digits are even and do not repeat.

Problem 3

Mystery Number

0. ___ ___ ___

Problem 4

Clue

The digit in the hundreds place is equal to the sum of the digits in the ones and tens places.

Problem 3

Clue

The number has no repeated digits.

Problem 4

Clue

The digit in the thousands place is 2 times the digit in the ones place.

Problem 3

Clue

All of the digits in the number are even.

Problem 4

Clue

The digit in the ten thousands place is 2 times the digit in the tens place.

Problem 3

Clue

The digit in the thousandths place is 2 times the digit in the hundredths place.

Problem 4

Clue

The digit in the hundred thousands place is 2 times the digit in the hundreds place.

Problem 3

Clue

The digit in the hundredths place is 2 times the digit in the tenths place.

Problem 4

Clue

There are two solutions to this problem.

Problem 3

Mystery Number

0. ___ ___ ___

Problem 5

Distinguishing Digits Cards

Clue

The number has no repeated digits.

Problem 5

Clue

The number has no repeated digits.

Problem 6

Clue

Two of the digits are odd.

Problem 5

Clue

All of the digits are even and positive.

Problem 6

Clue

The digit in the hundredths place is 3 times the digit in the tenths place.

Problem 5

Clue

The digit in the tenths place is less than the digit in the thousandths place.

Problem 6

Clue

The digit in the thousandths place is 2 times the digit in the tenths place.

Problem 5

Clue

The sum of all the digits is greater than 16.

Problem 6

Clue

There are two possible solutions to this problem. Can you find them both?

Problem 5

Clue

There are two possible solutions to this problem. Can you find them all?

Problem 6

Mystery Number

0. _____ _____ _____

Problem 6

Mystery Number

0. _____ _____ _____ _____

Problem 7

Clue

?
?
?

The digit in the tenths place is odd.

Problem 7

Mystery Number

___ ___ . ___ ___

Problem 8

Clue

?
?
?

The digit in the ten thousandths place is the sum of the digits in the tenths, hundredths, and thousandths place.

Problem 7

Clue

The digits in the tens and tenths places are the same, odd digit.

Problem 8

Clue

?
?
?

The number has no repeated digits.

Problem 7

Clue

The digit in the ones place is the sum of the digits in the hundreds, tens, tenths, and hundredths place.

Problem 8

Clue

?
?
?

The digit in the ten thousandths place is 3 times the digit in the tenths place.

Problem 7

Clue

The digit in the hundreds place is 2 times the digit in the tens place.

Problem 8

Clue

?
?
?

The digit in the hundredths place is a multiple of the digit in the tenths place.

Problem 7

Clue

The digits to the right of the decimal point are consecutive (like 3 and 4 or 8 and 9).

Problem 8

Clue

?
?
?

All of the digits are greater than 1, except for the digit in the thousandths place.

Problem 7

Clue

The digits in the hundreds and hundredths places are the same, even digits.

Problem 8

Distinguishing Digits Cards

Clue

The digit in the ones place is 5 more than the digit in the tenths place.

Problem 8

Clue

The digit in the hundreds place has consecutive factors whose sum equals the digit in the thousands place.

Problem 9

Mystery Number

_ _ _ , _ _ _ . _ _ _

Problem 9

Mystery Number

_ _ _ , _ _ _ . _ _ _ _

Problem 10

Clue

The number has no repeated digits.

Problem 9

Clue

None of the odd digits repeats.

Problem 10

Clue

The digit in the hundreds place is 1 more than the digit in the thousands place.

Problem 9

Clue

The digits in the ten thousands, thousands, and tenths places are the same.

Problem 10

Clue

The digits in the tens, ones, hundredths, and thousandths places are in order, beginning with 1 in the tens place.

Problem 9

Clue

The hundreds digit is 1 more than the thousands digit.

Problem 10

Clue

The sum of the digits in the tens and tenths places is 1.

Problem 9

Clue

The digit in the ten thousandths place is the sum of the digits in the thousands and ten thousands places.

Problem 10

Clue

The digit in the hundred thousandths place is the sum of the digits in the tenths and hundredths places.

Problem 10

Clue

The part of the number to the right of the decimal has no repeated digits.

Problem 10

Clue

The only odd digits appear in the thousandths, hundreds, and ones places.

Problem 10

Clue

The digit in the ones place is the sum of the digits in the tens and hundreds places.

Problem 10

Clue

The digits in the tens and hundredths places are the same and are half of the digit in the tenths place.

Problem 10

Clue

The digits in the tenths and thousandths places have exactly 3 factors.

Problem 10

base ten number system (page 52d) The base ten number system is the common number system we use. Our number system is based on the number 10 because we have ten fingers with which to group. With the common understanding that each group represents ten of the previous groups, we can write numbers efficiently. By extending the place-value system to include places that represent fractions with 10 or powers of 10 in the denominator, we can easily represent very large and very small quantities. Below is a graphic representation of counting in the base ten number system.

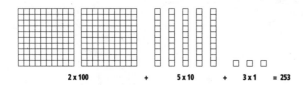

2 x 100   +   5 x 10   +   3 x 1   = 253

benchmark (page 23) A benchmark is a "nice" number that can be used to estimate the size of other numbers. For work with fractions, 0, $\frac{1}{2}$, and 1 are good benchmarks. We often estimate fractions or decimals with benchmarks because it is easier to do arithmetic with them, and estimates often give enough accuracy for the situation. For example, many fractions and decimals—such as $\frac{37}{50}$, $\frac{5}{8}$, 0.43, and 0.55—can be thought of as being close to $\frac{1}{2}$. You might say $\frac{5}{8}$ is between $\frac{1}{2}$ and 1 but closer to $\frac{1}{2}$, so you can estimate $\frac{5}{8}$ to be about $\frac{1}{2}$. We also use benchmarks to help compare fractions. For example, we could say that $\frac{5}{8}$ is larger than 0.43 because $\frac{5}{8}$ is larger than $\frac{1}{2}$ and 0.43 is smaller than $\frac{1}{2}$.

decimal (page 41) A decimal, or decimal fraction, is a special form of a fraction. Decimals are based on the base ten place-value system. To write numbers as decimals, we use only 10 and powers of 10 as denominators. Writing fractions in this way saves us from writing the denominators because they are understood. When we write $\frac{375}{1000}$ as a decimal— 0.375—the denominator of 1000 is understood.

The digits to the left of the decimal point show whole units, and the digits to the right of the decimal point show a portions of a whole unit. The diagram below shows the place value for each digit of the number 5620.301.

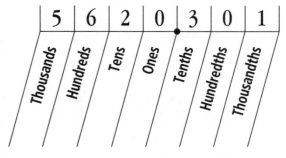

denominator (page 12) The denominator is the number written below the line in a fraction. In the fraction $\frac{3}{4}$, 4 is the denominator. In the part-whole interpretation of fractions, the denominator shows the number of equal-size parts into which the whole has been split.

equivalent fractions (page 20) Equivalent fractions are equal in value but have different numerators and denominators. For example, $\frac{2}{3}$ and $\frac{14}{21}$ are equivalent fractions. The shaded part of this rectangle represents both $\frac{2}{3}$ and $\frac{14}{21}$.

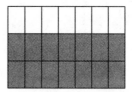

fraction (page 3) A number (quantity) of the form $\frac{a}{b}$ where a and b are whole numbers. A fraction can indicate a part of a whole object or set, a ratio of two quantities, or a division. For the picture below, the fraction $\frac{3}{4}$ shows the part of the rectangle that is shaded: the denominator indicates the number of equal-size pieces, and the numerator indicates the number of pieces that are shaded.

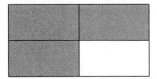

The fraction $\frac{3}{4}$ could also represent three of a group of four items meeting a particular criteria; the ratio 3 to 4 (for example, when 12 students enjoyed a particular activity and 16 students did not); or the amount of pizza each person receives when three pizzas are shared equally among four people, which would be $3 \div 4$ or $\frac{3}{4}$ of a pizza.

numerator (page 12) The numerator is the number written above the line in a fraction. In the fraction $\frac{5}{8}$, 5 is the numerator. When you interpret fractions as a part of a whole, the numerator tells the number of parts in the whole.

percent (page 67) Percent means "out of 100." A percent is a special decimal fraction in which the denominator is 100. When we write 68%, we mean 68 out of 100, $\frac{68}{100}$, or 0.68. We write the percent sign (%) after a number to indicate percent. (Some say that the percent sign % is what was left after people were writing a denominator of 100 many times and got sloppy.) The shaded part of this square is 68%.

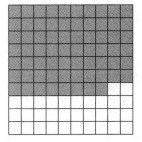

unit fraction (page 20) A unit fraction is a fraction with a numerator of 1. For example, in the unit fraction $\frac{1}{13}$, the part-whole interpretation of fractions tells us that the whole has been split into 13 equal-size parts and that the fraction represents 1 of those parts.